Material Interfaces as Energy Solutions

Topical Conference at the 2021 AIChE Annual Meeting

Boston, Massachusetts, USA and Online
7 - 11 November and 15 - 19 November 2021

ISBN: 978-1-7138-5691-7

Printed from e-media with permission by:

Curran Associates, Inc.
57 Morehouse Lane
Red Hook, NY 12571

AIChE
120 Wall Street, FL 23
New York, NY 10005-4020

Phone: (800) 242-4363
Fax: (203) 775-5177

www.aiche.org

Additional copies of this publication are available from:

Curran Associates, Inc.
57 Morehouse Lane
Red Hook, NY 12571 USA
Phone: 845-758-0400
Fax: 845-758-2633
Email: curran@proceedings.com
Web: www.proceedings.com

TABLE OF CONTENTS

623495 Delineating and Controlling the Electrode-Electrolyte Interfacial Reactions in High Energy Density Batteries

Abstract:

Rapid increase in global energy use and growing environmental concerns have prompted the development of clean, sustainable, alternative energy technologies. Renewable energy sources like solar and wind are a promising solution, but electrical energy storage (EES) is critical to efficiently utilize electricity produced from renewable sources as they are intermittent. EES is also the only viable near-term option for electrification of transportation sector. Rechargeable batteries are prime candidates for EES, but their widespread adoption for electric vehicles and grid electricity storage requires optimization of cost, cycle life, safety, energy density, power density, and environmental impact, all of which are directly linked to severe component materials challenges and the interfacial reactions between them. This presentation will focus on delineating and controlling the electrode-electrolyte interfacial reactions and consequent crossover of chemical species in high energy density batteries.

With an aim to develop sustainable battery technologies at an affordable cost with high energy density and free from supply chain issues, the presentation will focus on interfacial reactions in two battery systems: (i) lithium-ion batteries with low- or cobalt-free, high-nickel cathodes and graphite or lithium-metal anodes and (ii) lithium-sulfur batteries with sulfur cathode and lithium-metal anode or anode-free cells with Li_2S cathode. The high-nickel cathodes are synthesized with a coprecipitation of transition-metal hydroxide precursors followed by calcining with LiOH under controlled temperature and oxygen flow in order to obtain commercially-relevant particle size and morphologies. Practically-relevant pouch lithium-ion cells or lithium-sulfur cells fabricated with the cathodes and anodes are disassembled and characterized after hundreds or thousands of cycles with advanced characterization techniques, such as X-ray photoelectron spectroscopy, time-of-flight secondary ion mass spectrometry, and high resolution transmission electron microscopy, to track the anode-electrolyte interfacial reactions, cathode-electrolyte interfacial reactions, as well as the crossover of chemical species between the two electrodes through the electrolyte/separator during cycling. Based on the data, the impact of interfacial reactions and crossover of chemical species between the two electrodes on cell performance is established. The understanding gained is utilized to condition the electrode surfaces or modify electrolyte compositions in order to minimize undesirable interfacial reactions and enhance the cycle life of lithium-ion and lithium-sulfur cells.

623899 Investigating Li-Ion Behavior in ALD Coated NMC Cathode Materials Via Molecular Dynamics

Abstract:

High nickel content $LiNi_{1-y-z}Mn_yCo_zO_2$ (NMC) has emerged as a cathode capable of delivering high capacities in Li-ion batteries. In particular, high-nickel NMCs are being used commercially in electric vehicles (EVs) and are projected to comprise over a third of EV cathode chemistries by 2026.[1] However, high-nickel cathodes experience structural instability with the charge/discharge cycling of the battery, which leads to large capacity fading, voltage decay, and low rate capability of the overall battery. Atomic layer deposited (ALD) lithium-containing nano-coatings of NMC cathodes have been shown to help stabilize the transition metal sites while still allowing the movement of Li-ions across the coating, leading to greater capacity retention of the Li-ion battery.[2]

Our prior experimental work has demonstrated that alumina particle ALD produces a less than two nanometer, mixed $Al_2O_3 - LiAlO_2$ film that preferentially covers the transition metals of the cathode powder while leaving lithium relatively uncovered and presumably available for transport. Expanding upon this work, molecular dynamics simulations were used to investigate the behavior of surface lithium of uncoated cathodes and ALD coated cathodes. The Interface Force Field in the 12-6 Lennard Jones form was used to model the pairwise interactions.[3] Both amorphous and crystalline films were simulated at 393 K, with the crystalline film being used as a comparison to the amorphous, ALD-like one. The thickness and uniformity of the ALD coating affects the surface lithium movement. For example, lithium is able to completely travel through one simulated Al_2O_3 layer, while it cannot for two or more simulated Al_2O_3 layers. Further, the energy of the interface between the cathode and coating continues to decrease (i.e., become more favorable) as the simulated coating thickness increases until the energy plateaus off and is about the same for all later coating thicknesses. At this energy plateau, the representative energy of deposition for Al_2O_3 on $LiCoO_2$ is -1760 mJ/m^2 for the amorphous film and -2180 mJ/m^2 for the crystalline film.

By 2030, electric vehicles are projected to comprise almost 30% of new car sales globally. A protective nano-coating on the cathode that preserves battery capacity should help hasten the global transition from combustion vehicles to EVs.

References:

(1) Frith, James. Battery Prices and Manufactured Costs (Highlights from Bloomberg NEF EVO Report 2020), 2021.

(2) Hoskins, A. L.; McNeary, W. W.; Millican, S. L.; Gossett, T. A.; Lai, A.; Gao, Y.; Liang, X.; Musgrave, C. B.; Weimer, A. W. Nonuniform Growth of Sub-2 Nanometer Atomic Layer Deposited Alumina Films on Lithium Nickel Manganese Cobalt Oxide Cathode Battery Materials. *ACS Appl. Nano Mater.* **2019**, *2* (11), 6989–6997. https://doi.org/10.1021/acsanm.9b01490.

(3) Dhakane, A.; Varshney, V.; Liu, J.; Heinz, H.; Jain, A. Molecular Dynamics Simulations of Separator-Cathode Interfacial Thermal Transport in a Li-Ion Cell. *Surf. Interfaces* **2020**, *21*, 100674. https://doi.org/10.1016/j.surfin.2020.100674.

632863 Reserve Lithium-Ion Batteries for Lithium-Ion Free Cathodes

Abstract:

Lithium-ion Batteries' (LIBs) influence on every aspect of human lives has been unimaginable. One sector, which has received a tremendous boost because of LIBs is transportation. Electric vehicle (EV) sales have increased almost 20 folds from 2011 until now. This has caused a huge increase in the demand for LIBs across the world. Presently, there is heavy dependence on cathodes like $LiFePO_4$, $LiNiO_2$, $LiCoO_2$, and $LiNi_xMn_yCo_zO_2$ to meet energy expectations. Compounds without lithium in their lattice rarely have been used as cathodes either due to inability for reverse reactions or require lithium electrode, which causes lithium dendrites on continuous cycling. This has resulted in fewer efforts towards the development of such lithium-free cathodes. Some of these materials are ZnO, TiO_2, Cr_3O_4, MnO_2, V_2O_5, etc. Although these materials possess high theoretical capacities, many of them have a low potential for faradic reactions and have the risk of reducing to metal states during the electrochemical reactions. Hence, for such materials to commercialize, few measures have been taken namely, pre-lithiation of the cathode, film-forming additives, lithium additives, blended cathodes, etc. Some strategies like pre-lithiation are not practical and feasible, even though they help in overcoming solid electrolyte layer interface (SEI). Other strategies alter the cell voltage characteristics due to the presence of voids in the electrode.

We present a pioneering configuration called reserve lithium-ion batteries for in situ lithiation of the lithium-free electrodes with the use of lithium-electrode as a reservoir. For the initial formation of the SEI and lithiation of the MCMB electrode, a reservoir lithium electrode was brought into the cell circuitry. V_2O_5 cathode with meso carbon microbeads (MCMB) was used to fabricate high specific capacity and energy-dense reserved lithium-ion batteries (RLIBs). The RLIB system delivered a high charge specific capacity of 245 mAh g^{-1} at 0.2C rate. The tri-electrode system provided an initial high energy density of >880 Wh kg^{-1}. A parallel comparison of electrochemical performance between V_2O_5 half-cell, V_2O_5-MCMB full-cell, and V_2O_5-MCMB RLIB was conducted and observed. There is a distinct resemblance in the performance of RLIB performance with half-cell profile and in terms of capacity versus cycle number. The pouch-cell configurations of this work yielded a high energy-dense system, which cycled stably over more than 200 cycles at 0.5C rate. Using this innovative concept of reservoir charging would be of great importance for applications in various sectors of the society viz., space exploration, UPS, healthcare, EVs, grid storage, etc.

633516 Opening Remarks By the Chair/Co-Chair

Abstract:

The Chair/Co-Chair will introduce the session on Plasma Catalysis

628891 DFT-Informed Energetics of Plasma-Enabled Reactions Pathways and Microkinetic Modeling for Ammonia Synthesis on Transition and Low-Melting Point Metals

Abstract:

Ammonia is a crucial chemical (> 200 millions tons a year) due to its widespread use in fertilizer production, and its potential as a hydrogen carrier in a future "hydrogen economy." Currently synthesized using thermal catalysis, ammonia synthesis is currently a fossil energy-powered, high temperature-high pressure, energy-intensive process that is responsible for ~2% of the world's CO2 emissions. Ideally, as the world transitions to cleaner, renewable energy, the net CO2 emissions with ammonia synthesis would decrease. However, the harsh conditions required for thermally catalyzed ammonia synthesis demands a highly centralized, continuously-on process that is incompatible with renewable energy. Thus, currently, alternative methods to synthesize ammonia under milder conditions are actively sought, including plasma catalysis in plasma reactors.

Optimizing plasma-assisted ammonia synthesis, including optimizing the catalyst to be loaded on the reactor, is currently challenging due to an incomplete understanding of the most relevant reaction mechanisms involved in the process. One snippet of this debate include the role of vibrational versus radical species on ammonia formation. Some first principles-informed models have attempted to shed light on the relevant mechanism, but centered on vibrational species and only the typical Langmuir-Hinshelwood reactions associated with thermal catalysis were considered [1]. More comprehensive models including plasma radicals also attempted similar insights, but used generic, metal-agnostic parameters, and while including Eley-Rideal reactions, it only considered NH_X species [2]. These models thus may be at odds with experiments reporting the appearance of N2Hx species on the metal [3].

To enable first principles-informed comprehensive modeling of all plausible Langmuir-Hinshelwood (LH) and Eley-Rideal (ER) reactions involving N_YH_X surface species and plasma radicals we performed density functional theory (DFT) to obtain reaction energies for 50+ reaction events across 9 metal surfaces of different degrees of nitrophilicity (Fe, Co, Ni, Pd, Cu, Au, Ag, Ga, Sn), and used scaling relationships (both from the literature and new ones generated from our own select DFT calculations) to provide corresponding activation energies. Hydrogen dissolution events were also included in our calculations, motivated by previous observation of a hydrogen sink effect in experiments by some of us [4]. Ultimately, the obtained data was input into a microkinetic model, which we used to explore the effect of plasma composition on catalyst state and dominant reaction pathways. Our results indicate that plasma radical species such as N(g) and H(g) only need minimal concentration in the plasma phase (even several orders of magnitude lower than that of vibrational species) to make ER pathways dominant and the formation of N_YH_X species plausible. However, N_YH_X seem more likely to hydrogenate until saturation and spontaneous dissociation via ER reactions with H(g) radicals than to dissociate early via LH reactions. The dominant role of ER reactions makes for a crucial factor blurring differences in activities across different metals. Finally, we present relationships between energetic descriptors and experimentally measured catalyst activities in DBD and RF reactors.

1. Metha et al. Nature Catal. 2018, 1, 269-265

2. Van't Veer et al. *J. Phys. Chem. C* 2020, 124, 42, 22871–22883

3. Lea et al. *ACS Catal.* 2020, 10, 24, 14763–14774

4. Shah et al. ChemCatChem 2020, 12, 1200-1211

626387 Plasma-Assisted Catalytic Approaches for the Production of Chemicals

Abstract:

Plasma-assisted catalysis has grown rapidly in the past few years as a nonthermal, ambient pressure approach to activate stable molecules (e.g., N_2). Utilizing renewable energy, plasma processes have the potential to reduce the CO_2 footprint compared to thermal processes, and convert reactants at point sources owing to the modular nature of these systems. Our research in the area of plasma-assisted catalysis is focused on the ability to control the catalyst performance by tuning the plasma environment or plasma-catalyst interactions. We have been particularly interested in the use of nonthermal plasmas for C-H and N_2 activation via dry reforming of methane and ammonia synthesis reactions, respectively. In recent publications from our groups, we observed remarkable synergy between the plasma and catalyst for NH_3 synthesis from N_2/H_2: 1) the rate enhancements with plasma stimulation were more sensitive to N_2 concentration, 2) the ideal catalyst was not the same in a plasma environment as in traditional thermal catalysis, and 3) NH_3 yields above the thermal equilibrium limits were observed and sensitive to the catalyst composition. Further advances in plasma-assisted catalysis requires the fundamental knowledge of plasma interactions with catalyst surfaces. However, the consequences of plasma activation for the interaction of gas species with a catalyst surface are currently speculative, and observation of these species on the surface is very challenging. This presentation will focus on exploiting catalyst-plasma interactions for nitrogen and hydrocarbon activation, and the role of the plasma in enhancing catalytic performance.

627791 Models and Observations of Plasma-Catalytic Nitrogen Oxidation

Abstract:

Coupled plasma and catalyst enables reaction yields that exceed the contributions of the individual components. Reliable kinetic models are essential to rationalize observations and to guide the identification of appropriate plasma-catalyst combinations. Kinetic models of plasma- and of heterogeneous-catalytic-chemistries are individually well developed. Models to couple the two are still primitive. Here, we demonstrate DFT-parametrized microkinetic models to couple plasma and surface reactions. The models are applied to N_2 oxidation, an alternative nitrogen fixation process, enabled only with coupled non-thermal plasma and catalysts. We compare the intrinsic activity of Pt and Au under plasma conditions. We build integral reactor models coupling simplified plasma reaction and surface reaction networks to predict product yield. We first show plasma excited species can enhance turnover frequencies of both Pt and Au. The enhancing potential of catalysts to chemical yield from plasma, however, are condition- and material- dependent. Plasma/Pt generates more or comparable nitric oxide than plasma/Au under most thermal and plasma parametric conditions. Plasma/Au outperforms plasma/Pt under low thermal but high plasma excitations. Results are compared against experimental observations, highlighting that material selection must consider plasma properties, thermal conditions and reactor configurations.

630666 Methane-Products Process Intensification through a Nanosecond Plasma Discharge

Abstract:

The energy industry, such as other sections, moves toward more efficient systems and newer technologies hires by companies. Due to excellent energy-engineering setups that we have developed in many years, we could revolute the energy industry in the 20th century. The usages of natural gas, a clean feedstock, increase every year, and the majority of people in civilized countries have decided to move toward less polluted earth. Among all products that are produced at different plants, hydrogen plays a vital role. Producing hydrogen more efficiently at the existing plants is crucial. Moreover, hydrogen is a clean energy source and can help move toward a sustainable energy economy.

Most of (about 95%) hydrogen is produced from fossil fuels by steam reforming of natural gas, partial oxidation of methane, and coal gasification as of 2020. All these methods of hydrogen production require a considerable amount of energy due to high temperate operations. Therefore, the first question is how to produce hydrogen and light hydrocarbons cheaper with more eco-friendly technologies.

In this work, a nanosecond pulsed plasma reactor was designed to produce hydrogen and liquid hydrocarbons. The conversion rate, efficiency, and cost of the designed-made reactor show a promising method for converting hydrogen. Currently, hydrogen is priced between $3 and $8 per kilogram. We are presently producing hydrogen below $1.

Most of the researches that have been done on converting methane to hydrogen or other products have focused on the reactor's efficiency. This study attempts to present a novel reactor design and perform process intensifications for future industrial uses. Moving toward hydrogen as a source of energy can diversify our energy sources and enhance energy resiliency and security. Moreover, to avoid constructing gigantic plants to make them profitable, new-more efficient technologies are required to make the production in smaller-scale plants profitable. Also, having more distributed-smaller production plants would create opportunities to reach stranded reservoirs.

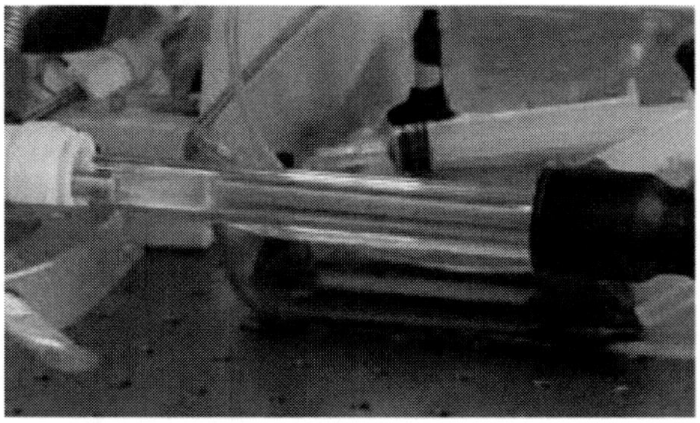

631226 Plasma-Assisted Upgrading of Methane: Mechanistic Insights from in-Situ PM-IRAS and OES Spectroscopy

Abstract:

Non-thermal plasmas (NTPs) produce reactive chemical environments including electrons, ions, radicals, and vibrationally-excited molecules and drive chemistry at temperatures at which such species are thermally inaccessible. The integration of a catalyst with reactive NTPs can promote novel chemical transformations. In particular, we are interested in upgrading of light hydrocarbons (e.g. methane) by direct coupling with nitrogen to produce value-added liquid chemicals (e.g. pyrrole and pyridine) in a plasma-assisted catalytic process. To design effective catalysts and plasma-catalytic processes, we need comprehensive understanding of the plasma catalytic processes including characterization of plasma-catalyst interactions. However, to date, the fundamental understanding of plasma-catalyst interactions is still in its infancy. In this work, we use a newly developed multi-modal polarization-modulation infrared-reflection absorption spectroscopy (PM-IRAS) tool with optical emission spectroscopy (OES) to probe plasma-catalyst interactions of CH_4/N_2 plasma species with different model catalytic surfaces (Ni, Pd, and Au). CH_4/N_2 feed gas mixture is introduced in three different ways: (1) CH_4-N_2 in sequence, (2) N_2-CH_4 in sequence, and (3) simultaneous with varying CH_4/N_2 ratio. We characterize excited species in the plasma-phase using OES, detect surface-adsorbed species using PM-IRAS, and detect gas-phase products using mass spectrometry. We find that different sequences in the feed gas introduction result in different surface-adsorbed species, and the presence of different surfaces dictates the surface-adsorbed species present. We also find that varying CH_4/N_2 feed ratio affects the macroscopic properties of the plasma and the production of intermediates such as cyanide (CN).

628510 Computational Evaluation of Plasmonic Photocatalysis in Au and Ag Nanoparticles

Abstract:

Many industrial chemical processes require high operating temperatures to drive catalytic reactions which are crucial to human development. These high temperatures are both energy demanding and detrimental to the long term life of the catalyst. Plasmonic metal nanoparticles offer an interesting alternative to traditional heterogenous catalytic processes due to their ability to be activated by light. One potential mechanism of action is the localized surface plasmon resonance, which involves oscillations in surface electron density that decay into so called "hot" electrons. These hot electrons can then transfer to unoccupied orbitals of surface intermediates to break chemical bonds or accelerate reactions. While plasmonic photocatalysis is a well-known event, the exact mechanism of these reactions is still debated. Understanding the precise workings of plasmon transfer is important for rational design of novel catalytic structures. Here, we utilize real-time, time-dependent density functional theory (RT-TD-DFT) to excite a system with oscillating electric fields and track the subsequent excited state dynamics in real time. This method models light of varying frequency in order to observe the mechanism of plasmon-mediated reactions in Au and Ag nanoparticles. Our results indicate that adsorbates can have qualitatively different charge transfer behavior. Specifically, O_2 loses electrons while N_2 gains electrons when exposed to identical electric fields on both Au and Ag nanoparticles. Additionally, charge density calculations indicate that these excitations are $\pi \rightarrow \pi^*$ on short time scales and a mixture of $\pi, \sigma \rightarrow \pi^*, \sigma^*$ as time evolves. We intend to further explain the differences between O_2 and N_2 via careful inspection of the electronic structure.

633459 First-Principles Understanding and Design of Metal Oxides for Photocatalytic Water Splitting

Abstract:

Renewable sources of energy are increasingly needed and solar production of hydrogen fuel from water offers significant potential to contribute to these needs if new photocatalyst materials and/or more efficient photocatalytic systems can be identified for visible-light-driven water photoelectrolysis. A variety of materials have been synthesized and tested for use as electrocatalysts. Among others, transition metal oxides are considered as one of the most promising electrocatalysts as they are earth abundant, eco-friendly, and stable in aqueous solution. Particularly, Bismuth Vanadate (BiVO4) has recently been identified as an inexpensive, promising photoanode for use in water-splitting photoelectrochemical cells (PECs). However, pristine BiVO4 exhibits poor electron-hole separation and sluggish water-oxidation kinetics, limiting its solar energy conversion efficiency. Significant research efforts have been made to enhance the PEC performance of BiVO4 through doping with metal/nonmetal elements and loading of oxygen evolution co-catalysts. However, the design and successful realization of commercially viable BiVO4-based photoanodes are still rather elusive. This talk will highlight the strong effects of crystal phase and doping on the photocatalytic performance of BiVO4 that have been found from combined first-principles and experimental studies. The improved understanding offers important guidance for the rational design of metal oxide-based photocatalysts.

632359 Structural and Stability Trends in Single (ABO_3) Perovskite Oxides from DFT-Optimized Bond Valence Structures

Abstract:

Ternary perovskite oxides (ABO_3) are a diverse class of technically important materials that exhibit exceptional properties for a variety of applications; as such, ternary perovskite material properties have been characterized in several high-throughput density functional theory (DFT) investigations. To ensure computational tractability, however, these previous investigations employ structural and stability approximations that are not rigorously supported. Neglecting these assumptions, we report a comprehensive investigation of 200 experimental ABO_3 compositions generated in all possible ternary perovskite geometries— as dictated by the 11 unique Glazer tilts— and optimized with DFT. Glazer tilt enumeration was performed with the Structure Prediction and Diagnostic Software (SPuDS) optimization package— which uses a bond valence method (BVM) based global instability index (GII) minimization procedure to predict octahedral (BO_6) tilting— and relaxed with GGA+U parameters compatible with the Materials Project (MP) database. We report a general linear correlation between GII and DFT energy that enables GII minimized structures to predict DFT optimized structures, particularly those that exhibit significant octahedral tilting. ~95% of DFT predicted ground states arise from the heavily distorted a-b+a- (orthorhombic) Glazer tilt, verifying that BO_6 distortions are the primary mode of stabilization in these compositions. Furthermore, we define the "energetic well depth" to be the difference between the undistorted a0a0a0 (cubic) geometry and the DFT predicted ground state, and show that linear scaling of the well depth energy predicts the decomposition enthalpy of the ground state— computed relative to MP tabulated competing phases— with RMSE = 40 meV/atom. By leveraging this relationship and slightly modifying the GII minimization procedure used by SPuDS to improve cubic lattice predictions, stability can be rapidly and accurately predicted from a single DFT optimization of the 5-atom cubic geometry. This study thereby constitutes the first time an integrated SPuDS+DFT scheme has been used to examine perovskite oxides, and represents a valuable addition to the growing arsenal of high-throughput investigative tools for targeted ABO_3 applications.

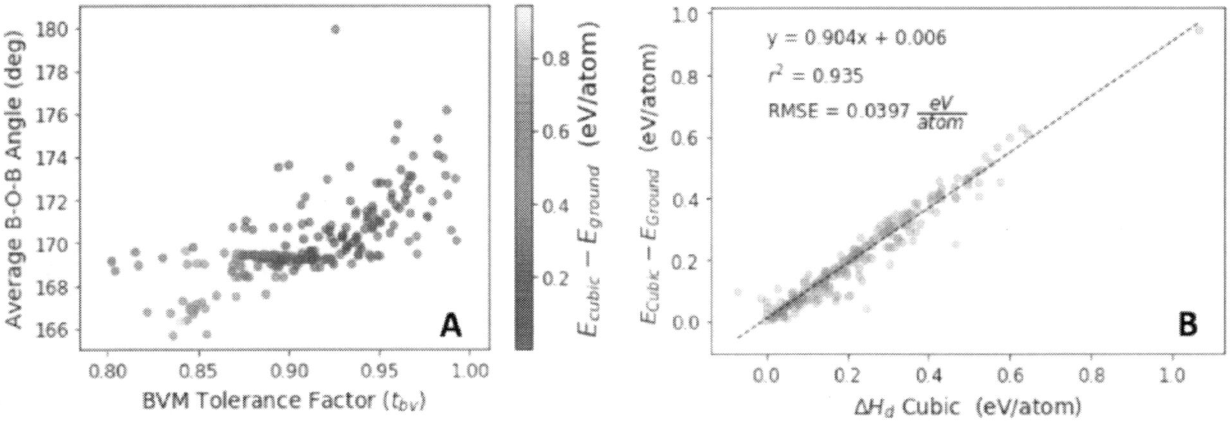

Shown for the 200 experimental compositions considered in our dataset: **A)** the relationship between the average B-O-B angle of the minimum DFT energy polymorph (i.e., the DFT predicted ground state, y_1-axis), the DFT energy difference between the undistorted cubic polymorph and the minimum DFT energy (distorted ground state, y_2-axis) polymorph, and the bond valence Goldschmidt tolerance factor (t_{bv}, x-axis). In the region $t_{bv} < 1.0$, the average B-O-B angle of the DFT ground state polymorph and the depth of the DFT predicted energetic well both increase as t_{bv} decreases, in agreement with SPuDS. In the region $t_{bv} > 1.0$, A and B site displacements supersede BO_6 tilting as the dominant form of distortion; this region is therefore beyond the scope of the current investigation. **B)** the stability of the cubic polymorph with respect to the convex hull (ΔH_d Cubic, x-axis) is linearly related to the depth of the energetic well (E_cubic - E_ground, y-axis). Linear regression shows that the magnitude of the energetic well depth is ~90% that of the cubic phase stability ($r^2 = 0.935$). Therefore, the DFT predicted ground state stability with respect to the Materials Project (MP) computed convex hull can be estimated as $\Delta H_{d,\ ground} = 0.1 \cdot \Delta H_{d,\ cubic}$ with reasonable accuracy (RMSE = 0.0397 eV/atom).

633322 Extracting the RedOx Thermodynamics of Perovskites via Combined Experiment and Theory

Abstract:

The analysis and design of thermochemical looping systems for renewable fuel production and energy storage requires detailed knowledge of the partial molar enthalpies and entropies of reduction of the active metal oxides. Experimental data normally consists of measured non-stoichiometries at a series of temperatures and oxygen partial pressures. Thermodynamic properties are then extracted via the van't Hoft or other analysis, such as the compound energy formula. These techniques often allow for trade off of enthalpy or entropy, providing non-unique solutions. In contrast computational methods such as density functional theory struggles to reproduce the complex temperature dependent behavior due to the large number of potential configurations. Here we present a new method for extracting thermodynamic properties which leverages a combination of DFT and experiment. This method removes the ambiguity between enthalpy and entropy, while providing temperature dependent behavior. The solution does not require pre-made assumptions.

633448 Materials Discovery and Development for Lower Temperature and Near Isothermal Thermochemical H2 Production

Abstract:

Materials discovery and development related to thermochemical materials for H2 production has historically been focused on materials that have large entropy changes and moderate enthalpies, in order that temperature swings between reduction and oxidation steps are minimized and thermal reduction temperatures are reduced below 1400 °C. Isothermal or near-isothermal cycling has been proposed as a means to decrease sensible heating requirements to the solid phase, but theoretical and experimental work has yielded mixed predictions and results. Based on our recent theoretical and experimental work, we postulate that the prior mixed results reported in literature stem from a combination of materials and conditions for isothermal operation (e.g., CeO2-δ at pO2 ~ 10-4 atm and 1400°C-1500 °C), that has been far from those desired for optimal performance. In this work, we implement a materials design concept that identifies and investigates new material classes, compositions, and operating conditions that are ideal for isothermal thermochemical H2 production. Starting from La-Sr-Mn perovskites, which have been theoretically and experimentally identified as having more favorable properties than ceria, we are utilizing high throughput screening coupled to direct DFT computations of down-selected materials for key performance properties such as oxide formation energies and oxygen vacancy formation and migration energies. DFT computation and screening is coupled closely with experimental synthesis of powdered, replica and direct foam compositions that afford enhanced heat and mass transfer properties. Compositions are experimentally characterized through thermogravimetric analysis and relaxation experiments at a larger scale to test the stability and performance of the replica and foamed compositions.

629134 Thermodynamic Characterization of Doped Spinels for Thermochemical Fuel Production

Abstract:

Technologies such as solar thermochemical hydrogen (STCH) production, which leverage the oxygen-exchange capacity of metal oxides to split water, are thermodynamically attractive, as the entire solar spectrum can be exploited to supply process heat. Here, a metal oxide is subjected to cyclic reduction-oxidation reactions that facilitate either oxygen or hydrogen production, depending on the operating conditions. Chemical thermodynamics suggest that the forward direction of each redox half-reaction is favorable in different temperature and/or oxygen partial pressure regions, and as a result, in practice, large irreversibilities generally arise from sensible heating between redox regimes. To address this issue, recent research has primarily focused on: (1) investigating novel perovskite materials with tunable thermodynamic properties that enable greater reduction extents at more moderate temperatures; and (2) integrating auxiliary techniques, such as mechanical or thermochemical pumping, that reduce the operating oxygen partial pressure (and thus temperature) in the endothermic reduction step. However, until highly effective and reliable solid-solid heat recovery methods are devised, inefficiencies inherent to temperature-swing operation will continue to negatively impact performance. Alternatively, at the expense of lower hydrogen productivity, the water-splitting (oxidation) step can be operated at temperatures where reduction is favorable, thereby minimizing thermal stresses and avoiding complexities associated with solid-solid heat recuperation. Herein, motivated by the immediate practicality of isothermal operation, the behavior of low-cost cobalt-iron aluminate spinels was characterized under conditions relevant to isothermal STCH. Using thermogravimetry, equilibrated changes in oxygen content were quantified at different oxygen partial pressures and temperatures to extract thermodynamic properties, namely the standard molar enthalpy and entropy of relevant defect reactions, and thus enable performance predictions. Results indicate that this class of redox intermediates is capable of higher hydrogen productivities under isothermal operation than previously established benchmark materials (e.g., ceria and lanthanum manganite-based perovskites) evaluated under temperature-swing. The favorable thermodynamic properties of spinels at conditions better suited for industrialization present a step forward towards sustainable hydrogen production with solar thermal energy.

622622 Sustainable High-Purity Nitrogen Production Via a Coupled PSA-Thermochemical Process for the Ammonia Industry

Abstract:

In order to achieve the Paris agreement's target of maximum global warming of 1.5 °C, a sustainable and CO_2 lean production, not only of fuels, but also of key commodities like ammonia, is necessary. Ammonia is a fundamental chemical commodity for fertilizer production, and becomes increasingly important as an energy vector, especially considered as fuel for maritime applications. Since not only the production of H_2, but also the production of the feedstock N_2 itself causes high CO_2 emissions, its sustainable production is mandatory. Economic analysis reveals that the combination of a solar-energy-powered pressure-swing absorption (PSA) and a redox-oxide-based, solar thermochemical cycle decreases the thermal energy input of high-purity N_2-production significantly. The viability of thermochemical air separation and the benefit of using pre-purified air has already been proven in laboratory scale experiments, but the approach of coupling PSA with such a thermochemical cycle has not yet been demonstrated.

In this contribution we present an overall approach including both the redox oxide material selection as well as the solar reactor design. The characteristics of the selected redox materials are analyzed by in-situ high-temperature x-ray diffraction (XRD) in a temperature range of 500-1100°C and thermogravimetric analysis (TGA). Various oxygen partial pressures (0-83 % O_2) are applied in both characterization methods.

The proposed air separation process is demonstrated with a PSA unit coupled to an indirectly heated, pressurized packed bed reactor (Fig. 1). The influence of the process parameters such as pressure, reduction and oxidation temperatures on the performance criteria are shown and compared to results gained by simulations. Moreover, we illustrate how the process parameters can be influenced by tuning the composition of a representative redox material $Ca_{1-x}Sr_xMnO_3$. Therefore, the impact of the Sr-amount on the structure, structural changes and the enthalpy are presented. The performance of the overall system is evaluated on the basis of the achieved purity of the outlet gases, the volume flow rates and the efficiency.

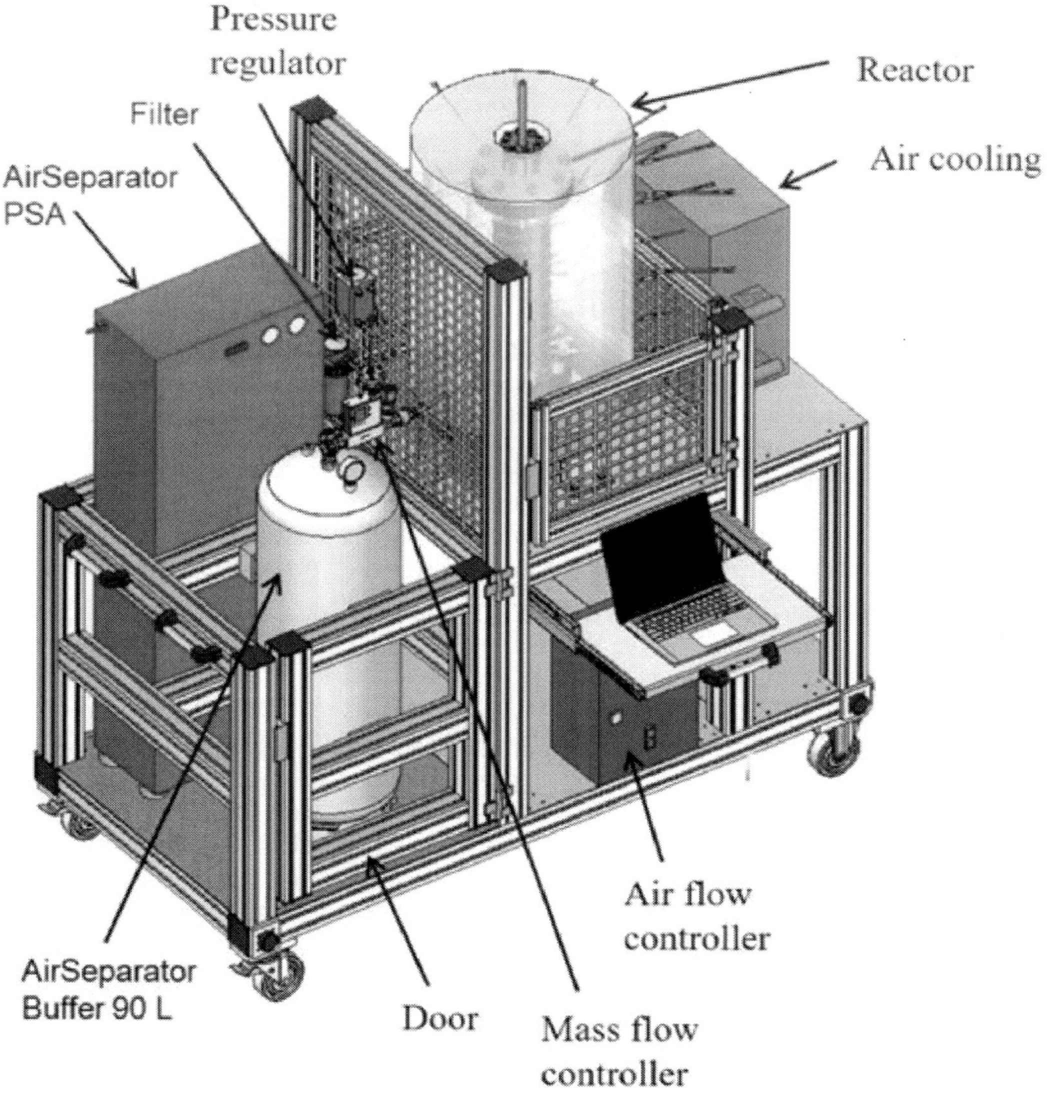

Fig. 1. Reactor design for a two-step air separation process

633022 High-Purity Nitrogen Production from Air By Pressure Swing Adsorption Combined with $SrFeO_3$ Redox Chemical Looping for Trace Oxygen Removal

Abstract:

This work focuses on the production of high purity nitrogen from air via a combination of pressure swing adsorption (PSA) with a downstream redox chemical looping cycle used as a deoxygenation unit. The non-stoichiometric perovskite $SrFeO_{3-\delta}$ was selected for the redox chemical looping cycle because of its favorable thermodynamics, rapid oxidation kinetics and intermediate reduction temperatures. A first principles analysis of the system showed that this combined process could improve upon both the energy balance and production capacity of a PSA system. The chemical stability of the material was demonstrated over 250 redox cycles via thermogravimetry with less than 1 % degradation in the chemical storage capacity. The oxidation kinetics were also measured at 523 K over a range of oxygen concentrations, and fit with a suitable kinetic model. The kinetics were then incorporated into a 1D convection-diffusion model of a packed bed reactor. The model indicates that, for a targeted oxygen impurity of $x_{O2} < 3 \times 10^{-6}$, a chemical looping unit added downstream to a PSA system could approximately triple the capacity and reduce the energy demand to 14 kJ/mol of N_2 [Bulfin *et al.* 2020]. The results are promising for further study, with the goal of bridging the gap between PSA and cryogenic systems in terms of scale, purity and cost.

Bulfin, B., Buttsworth, L., Lidor, A., & Steinfeld, A. (2020). High-purity nitrogen production from air by pressure swing adsorption combined with SrFeO3 redox chemical looping. Chemical Engineering Journal, 127734.

.

632564 Interlayer Vibrational and Thermal Transport Phenomena in 2D Layered Perovskites

Abstract:

2D lead halide perovskites offer a versatile platform for the creation of atomically precise interfaces between organic and inorganic layers. In this talk, I will describe the vibrational, thermal, and phase change properties of 2D alkylammonium lead halide perovskites. Highlights will include 1) selective order/disorder transitions of the organic subphase, 2) observation of zone-folded longitudinal acoustic (zf-LA) phonons in 2D superlattices, and 3) evidence for ballistic phonon transport in interlayer heat transfer.

634247 Boosting Hybrid 2D Perovskite Stability through Computational Screening of Ligand Chemistries

Abstract:

2D hybrid perovskite materials are attractive from the perspective of providing a much larger design space of organic ligand chemistries in comparison with bulk 3D materials. Likewise, many recent reports have highlighted how tuning the ligand chemistry can impact the opto-electronic properties and stability profiles of the resulting 2D materials. In this talk, we'll describe how we have implemented a high-throughput molecular simulation framework to screen prospective ligand chemistries for synthesizability and stabilization of the perovskite lattice. Using this framework, we have been able to characterize tens of thousands of perovskite materials and establish some novel design rules for ligand chemistries. We will also present a few case studies of materials that have been synthesized and tested in response to this effort.

625598 Engineering Entangled Photon Pairs with Metal–Organic Frameworks

Abstract:

The discovery and design of new materials with competitive optical frequency conversion efficiencies can accelerate the development of scalable photonic quantum technologies. Metal–organic framework (MOF) crystals without inversion symmetry have shown potential for these applications, given their nonlinear optical properties and the combinatorial number of possibilities for MOF self-assembly. In order to accelerate the discovery of MOF materials for quantum optical technologies, scalable computational assessment tools are needed. We develop a multi-scale methodology to study the wavefunction of entangled photon pairs generated by selected non-centrosymmetric MOF crystals *via* spontaneous parametric down-conversion (SPDC). Starting from an optimized crystal structure, we predict the shape of the $G^{(2)}$ intensity correlation function for coincidence detection of the entangled pairs, produced under conditions of collinear type-I phase matching. The effective nonlinearities and photon pair correlation times obtained are comparable to those available with inorganic crystal standards. Our work thus provides fundamental insights into the structure–property relationships for entangled photon generation with metal–organic frameworks, paving the way for the automated discovery of molecular materials for optical quantum technology.

634315 Invited Talk: A Charge Transfer Framework That Describes Supramolecular Interactions Governing Structure and Properties of 2D Perovskites

Abstract:

Organometal halide perovskites can efficiently absorb light; perovskite solar cells can generate electricity with power conversion efficiencies now exceeding 25%. However, halide perovskites decompose when exposed to light, heat, or moisture, making them unsuitable for use as active components in solar cells. Two-dimensional perovskites with organic ammonium spacer ligands offer improved environmental stability, but the power conversion efficiencies of the solar cells containing these two-dimensional perovskites are low, reflecting inefficient charge transport. As the interactions between organic ligands and the perovskite octahedra affect the octahedral tilts, directly affecting charge transport, there is an opportunity for a comprehensive molecular design strategy that can manipulate organic-inorganic interactions, control octahedral tilt, enhance charge transfer, and impact device performance. In this talk, I will discuss our approach to molecular design, and using both theoretical and experimental support, elucidate the role of the organic cation in controlling the macroscopic properties of the resulting two-dimensional perovskites. Importantly, we learned that attractive point interactions within the organic bilayer result in a withdrawal of the organic cation from the interstitial site, and less penetration of the organic cation correlates to better performing solar devices. This design paradigm can guide future innovations in solution processable perovskite optoelectronics, with the goal of creating both high performing and stable next generation solar cells.

627921 Molecular Modeling of Halide Diffusion in 2D Organic-Inorganic Hybrid Perovskites

Abstract:

Organic-inorganic hybrid perovskites are promising semiconductor materials for a wide range of optoelectronic applications due to their unique properties, including high carrier mobility, solution processibility, and strong light absorption. However, facile halide diffusion is a source of instability for 3D perovskites that leads to rapid degradation of device performance. In contrast, 2D perovskites exhibit relatively suppressed halide diffusion that is hypothetically further improvable via surface cation design. Nonetheless, the halide diffusion mechanism in 2D perovskites remains largely unknown and is challenging to experimentally characterize. To understand the microscopic origin of the diffusion and the role of organics cations, we have employed the molecular dynamics (MD) simulation to explore different halide migration paths at the atomic scale and the role of surface cation in suppressing diffusion. Steered MD and thermodynamic integration were used to characterize the free energy profiles of each migration path at surface/bulk layers for various systems. In comparison to 3D perovskites, we observe that 2D perovskites exhibit relatively high diffusion activation energies and a large the asymmetry across the accessible channels, which accounts for the suppression of halide diffusion. In addition, the change in the energy barrier when applying different organic cations and the number of inorganic layers offers guidelines for molecular engineering of perovskite materials toward long-term stability.

Polymer-guided interfacial assembly of nanoparticles into low-dimensional architectures

Gaurav Arya*, Yilong Zhou, Tsung-yeh Tang, and Brian Hyun-jong Lee

* Presenter, Department of Mechanical Engineering and Materials Science, Duke University, Durham, NC 27708, USA

ABSTRACT

In this talk, I will discuss how polymer-grafted nanoparticles at fluid-fluid interfaces can self-assemble into unique higher-order structure due to the three-way competition between interparticle interactions and the interfacial tensions of the nanoparticle with the two liquids. I will begin by demonstrating through molecular dynamics simulations the interfacial assembly of spherical nanoparticles into open clusters with tunable orientation and bond angles; semi-linear structures like serpentine strings and branched networks; and layered structures such as corrugated monolayers and square-ordered bilayers. Next, I will show how tailored Monte Carlo minimization methods along with analytical models of interparticle and interfacial interactions can be used to quickly map out the full repertoire of layered architectures achievable through this assembly approach. Lastly, I will show, also via molecular dynamics simulations, how interfaces can be used to orient shaped nanoparticles, especially those with faceted geometries whose subsequent assembly at the interface can yield an interesting class of porous architectures. Overall, this work demonstrates the power of polymer-guided, interfacial assembly for fabricating next-generation nanoparticle architectures with possible applications in plasmonics, optics, membranes, and catalysis where control over particle orientation and spatial arrangement is critical for function.

623511 Soft and Stretchable Energy Harvesting Using Metal/Gel Interfaces

Abstract:

Capacitors are widely used in electrical circuits due to their ability to store electrostatic energy and discharge current almost instantly when needed. Variable capacitors are those that can change capacitance – usually via geometric changes—to convert mechanical energy to electrical energy. Despite the headway, harvesting mechanical energy by varying capacitance is limited to sliding or oscillating the electrodes up and down in a systematic way due to their rigid nature. In addition, the relatively large gap between the electrodes in a traditional rigid capacitor results in low capacitance, limiting the charge storing capacity. To address these limitations, we utilize a so-called electric double layer capacitor. Such a capacitor forms at the interface between metal and electrolyte, due to the formation of oppositely charged layers form at the interface. Electrical double layers innately offer higher capacitance than conventional capacitors due to the small distance between the charged layers. We use liquid metal for the electrode, which allows the area of these interfaces to vary in response to mechanical forces. Thus, we were able to create variable area capacitors that can harvest mechanical energy. Since they are soft and deformable, they can convert all modes of mechanical input to electricity including pressing, stretching, bending, and twisting. These devices generate ~1 mW m^{-2} by harnessing energy from mechanical motion without the need for an external battery, unlike the conventional electrostatic energy harvesters. We have characterized the behavior of these devices as a function of several parameters including material properties, design architecture, and mechanical deformation. The devices perform as expected and the response of the devices to deformation matches a physics-based model. The liquid metal-based soft device generates an electrical signal when deformed and works underwater, which may be useful for oceanic energy harvesting as well as wind energy harvesting. Besides, these devices can harvest human motion thereby find applications in wearable electronics, healthcare systems like rehabilitation and prosthetics.

630815 Morphology and Charge Transport Predictions across Organic Photovoltaic Components Using Coarse-Grained Molecular Dynamics

Abstract:

Optimizing optical and electronic properties of organic photovoltaic active layers is important for realizing low-cost, high-efficiency solar power. Molecular simulations can guide selection of organic components and processing conditions if sufficiently large length scales and time scales can be accessed. We manage ensembles of multiscale molecular dynamics simulations of polymer electron-donor and both fullerene and non-fullerene electron-acceptor materials, making extensive use of MoSDeF tools signac, foyer, and mBuild. Coarse-grained representations and GPU acceleration are used to access long-time equilibration dynamics. Atomistic representations from the coarse models are used for calculating charge transfer dynamics between chromophores, in service of predicting charge mobility as a proxy for active layer quality. We find simplified coarse-grained representations to reliably predict experimentally-validated morphologies for broad classes of organic components and describe specifics for ITIC, PTB7, BDT-TPD, PCBM, and P3HT. Varying agreement between our predictions of charge mobility and measured experimental values reinforce charge assumptions of charge delocalization that are valid for small-monomer polymers, but which need improvement for large non-fullerene acceptors. The predictive capabilities of coarse-grained models suggest that as these simulations become more routine, adding non-equilibrium and rheological processing effects and improving the speed and accuracy of charge transport calculations will be the next frontiers in organic photovoltaic optimization.

631527 Exploring the Energy Landscape of Soft Glassy Systems

Abstract:

The dynamics and structure of systems beyond the glass and jamming transition have been of interest for a long time due to their importance in a wide range of fields. Molecular glasses are essential active layers in organic light-emitting diodes, while the phenomena of jamming is important in dense emulsions and granular flows. The slow dynamics and complexity of theses systems has restricted the ability to simulate them using conventional methods, particularly at low temperatures or high volume fractions. The properties of the low energy states are of particular importance to answering many fundamental questions surrounding the glass transition and in connecting simulation results to experiments. In this study, we use an energy landscape based approach to probe the low-energy inherent structures and provide an algorithm to drive the system to lower energy states. The algorithm is inspired by metadynamics, and proceeds as a high dimensional basin filing algorithm that allows us to explore the energy landscape of a model soft-sphere foam system. We demonstrate the ability of the algorithm to reach low energies while revealing interesting geometric characteristics about the nature of the landscape. We study how connected metabasins are organized relative to a starting energy minimum and their orientations relative to each other. Finally, we compare our method with conventionally used methods like simulated annealing and swap Monte Carlo and discuss the advantages and disadvantages of each.

632594 Optimized Generation of Initial Conformations for the Simulation of Amorphous Polymer Systems to Reduce Required Simulation Resources

Abstract:

Significant breakthroughs have been derived from the simulation of non-crystalline polymer phases including optimization of such polymers for mechanical, membrane, surface, biomedical and microelectronics applications. Unfortunately, these simulations require long simulation times when the molecular weight of interest is at or above the entanglement length of the polymer chain. Some reduction in the required simulation time is attainable through the use of simulation algorithms that efficiently sample conformation space, but an additional reduction can be obtained by producing an accurate initial conformation of the polymer system that is close to the equilibrated structure. This work investigates the effect of density on the initial conformation and the importance of long-range corrections in both cohesive energy and density. We utilize the Suter-Theodorou initial guess algorithm that employs a Rotational Isomeric States model for short range conformational choices coupled with long range self-avoiding corrections. However, we carry this out as a function of the density of the periodic cell. Depending on the nature of the molecular interactions (dispersion, polar, and hydrogen bonding), there appears to be an optimal density at which the initial conformation is generated. Generation of the initial guess at the experimental density can produce unrealistically high energy conformations. While generation at lower density can alleviate these high energy conformations, it also reduces the chain entanglement. This entanglement is significant because chain diffusion is slow, and insufficient entanglement requires extensive simulation time to recover. By manipulating density, we can optimize both energetics and entanglement to produce an initial conformation close to the equilibrium structure of the amorphous polymer. Additionally, we utilize an updated protocol for calculating the long-range energy and pressure corrections that extrapolates a series of reproduced periodic structures. This approach avoids the conditional convergence of the many of the commonly used methods for electrostatic energy correction.

632864 Adaptive Grid-Based Method for Mapping Cavity Connectivity in Thermal Crystals and Amorphous Materials

Abstract:

The cavities in ideal, zero Kelvin crystal structures are straightforwardly charactized by the largest sphere that can be placed in each void. For example, in an fcc crystal it is well-known that the largest sphere that can fit in an octahedral (O) site has a diameter 0.414 times the atomic diameter [1]. This method, however, does not account for the non-spherical shape of each void nor does it account for fluctuations in the size and shape of the cavity at finite temperatures. Additionally, it is not adaptable to amorphous materials in which the cavity centers are not easily identifiable.

We develop an adaptive grid-based method for computing the size and shape of cavities in thermal crystals and amorphous materials. We identify pathways connecting cavities by launching stochastic trajectories and compute the energy barrier for an absorbed atom to hop between connected cavities using the nudged elastic band (NEB) method. We calibrate our methodology by analyzing a zero Kelvin fcc crystal. The resultant cavity map has one O site and two tetrahedral (T) sites per atom, with each O site connected to eight nearest neighbor T sites. We demonstrate the sensitivity of our results to two adjustable parameters, the probe diameter and the number of trajectories. Finally, we present cavity maps for an ensemble of amorphous configurations for a binary Lennard-Jones mixture [2].

This work was performed under the auspices of the U.S. Department of Energy by Lawrence Livermore National Laboratory under Contract DE-AC52-07NA27344. Document release number LLNL-ABS-821430.

Figure. (a) An fcc unit cell showing O and T sites in blue and red, respectively. (b) The connectivity map for (a) showing that each O site is connected to 8 T sites. (c) An amorphous configuration showing six cavities. (d) The connectivity map for (c) showing how the cavities are connected.

References

1. Fukai, Yuh. The metal-hydrogen system: basic bulk properties. Vol. 21. Springer Science & Business Media, 2006.
2. Kob, Walter, and Hans C. Andersen. "Testing mode-coupling theory for a supercooled binary Lennard-Jones mixture I: The van Hove correlation function." *Physical Review E5* (1995): 4626.

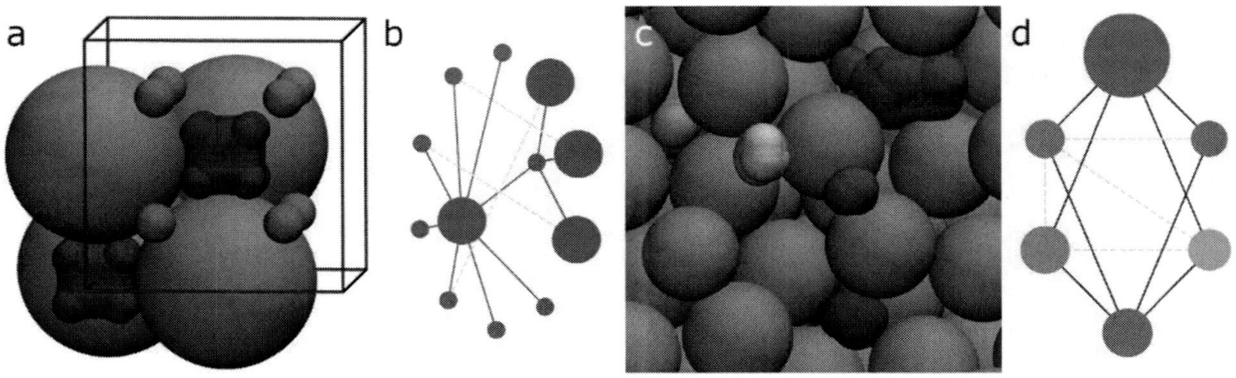

632638 *Electrosprayed Scalable 3D Graphene-CNT Electrodes for Li-Ion and Fuel Cell Applications*

Abstract:

Lithium-sulfur batteries (LSB) and advanced lithium-ion batteries (LIB) demonstrate promise as the next generation energy storage and conversion (ESC) technology especially as it pertains to wearable technology and electric vehicles. Although LIB dominate the battery market (>90%) due its reliability, long cycle life, and market maturity, alternative means of renewable energy must be discovered in order to reduce the cost of materials, dendritic issues at the solid-electrolyte interphase (SEI) layer, and limited charge capacity (~250 mAH h g^{-1}). The next generation of future batteries is the lithium-sulfur battery due its inexpensive cost ($0.25/kg sulfur), high theoretical energy density (2600 Wh kg^{-1}), and safety. How LIB and LSB systems incorporate into military applications such as drones and battery backs is what the interests the Army. However, LSB face many challenges before commercialization into wearable technology and electric vehicles due its fast capacity degradation, polysulfide "shuttling" effect, and short cycle life (<300 cycles). Carbon nanomaterials constitute an ideal miniaturized composite platform for integration with other solid-state electrode materials. Air-controlled electrospray has been recognized as an effective means of fabricating cathodes using a convective air jet flow that accelerates the drying and deposition process. This technique provides a uniform and well-dispersed coating onto the current collector while using low-cost commercial carbon nanomaterials in a matter of minutes. Currently, self-assembly is conducting using the bottom–up synthesis of 3D macrostructures using graphene and carbon nanotubes (CNTs) as building blocks. We demonstrate a novel, all-aqueous, and scalable 3D platform materials design process for the synthesis of carbon nanocomposite aerogels to serve as porous ultralightweight electrodes for the next generation energy storage and conversion applications. Specifically, the negatively charged surface functional groups on graphene oxide and oxidized CNTs will be the sites for electrostatic coordination of positively charged noble metal cations and polyelectrolytes from aqueous solutions. This enables the incorporation of carbon nanomaterials with any polyelectrolyte and noble metal nanostructures with precise connection of each of the individual components while maintaining their unique properties in both LIB and LSB systems. Through a scalable air-controlled electrospray approach, we can develop hierarchical porous 3D nanocomposite cathodes comprised of carbon nanomaterials with any precious or noble metal nanostructures (Pt, Pd, Au, Ru, Ag, and Ru) and polyelectrolyte that demonstrate a synergistic network between the active materials, carbon materials, and metal nanostructures. Electrochemical techniques such as cyclic voltammetry, linear sweep voltammetry, galvanostatic charge-discharge, and impedance spectroscopy are used to determine the electrical and ionic conductivity, electrocatalytic activity, and supercapacitor performance of the aerogels as lightweight high-power and energy density electrodes.

624456 Modeling the Brønsted Acidity of Lanthanum Exchanged Faujasite

Abstract:

Lanthanum stabilizes faujasite in fluid catalytic cracking, which increases catalyst material lifetimes. Despite its extensive use for decades, many unknowns remain on how lanthanum increases faujasite hydrothermal stability. The hydrothermal stability of zeolites decreases by dealumination, which occurs via reactions that involve water molecules. Brønsted acid sites give desired catalytic properties to zeolites, however they reduce hydrothermal stability by attracting water molecules to aluminum tetrahedra and promoting dealumination. The energetically favored binding sites of water molecules and lanthanum ions in hydrogen exchanged faujasite were determined with density functional theory (DFT) calculations and ab initio molecular dynamics (AIMD) simulations. Potentials of mean force of proton transfer were obtained with constrained AIMD simulations to quantify the Brønsted acidity of faujasite, with and without lanthanum ions. Results indicate that water and lanthanum ions are energetically favored to bind in the same sites in faujasite, and that lanthanum ions increase the acidity of Brønsted sites adjacent to, and not adjacent to, bound lanthanum ions in faujasite. Implications toward hydrothermal stability are discussed, since lanthanum ions change the protonation state, and therefore hydrophilicity, of aluminum tetrahedra in faujasite.

629492 Charge Transport Mechanism in Sodium Selenide (Na_2Se) for Sodium-Selenium Batteries: A First Principles Study

Abstract:

Sodium–selenium (Na–Se) batteries are recognized as promising energy storage systems because of their high theoretical volumetric capacity (~ 3250 mAh/cm^3), which is of particular importance for portable devices and electric vehicles (EVs) with the limited battery packing space. Sodium selenide (Na_2Se) is the final discharge product for the Na–Se batteries and is expected to limit the overall battery conductivity because of the insulating nature. To effectively utilize the active materials, it is necessary to have a better understanding of the charge transport mechanism in Na_2Se. Based on first-principles simulations, we first evaluate the contribution of potential charge carriers (ionic and electronic defects) to the conductivity in Na_2Se by calculating the formation energies and diffusion barriers. Negatively charged sodium vacancies (V_{Na}^-) and positive sodium interstitials (Na_i^+) are found to be the defects with the lowest formation energy (0.80 eV) in Na_2Se, followed by hole polarons (h_p^+) with a formation energy of 1.95 eV. Our simulations demonstrate that while the V_{Na}^- has a higher diffusion barrier than the h_p^+ (0.25 eV *vs.* 0.11 eV), its contribution to the conductivity is the most dominant, on the order of 10^{-15} S/cm, due to the high concentration. On the other hand, the electronic conductivity mediated by the h_p^+ is predicted to be negligible because of its low concentration. Once the vacancy–polaron ($V_{Na}^- - h_p^+$) complex forms, the energy cost to break the pair is predicted to be 0.78 eV, implying that the charge transport could be hindered due to its charge neutrality. Our calculations also illustrate that compressive strain (which is potentially induced by surrounding host materials) could reduce both ionic and electrical conductivity of Na_2Se. Lastly, we will discuss a possible strategy to modify the ionic and electronic conductivities in Na_2Se by controlling the strain.

624018 Nitrogen-Rich Free-Standing Carbon Cathode for Improved Lithium-Sulfur Battery Performance

Abstract:

A study on the production of free-standing carbon cathode is an attractive method to improve the low active material ratio, which is a problem with existing Li-S batteries. This work introduces a free-standing carbon material with superior porosity properties for enhanced Li-S battery performance. The free-standing carbon cathode synthesized in this study shows outstanding physical stability by showing superior elastic properties based on a porous structure. Additionally, the free-standing cathode made through electrospinning of $NaBH_4$ and polyacrylonitrile (PAN) has a high nitrogen atom equivalent to about 20%. This in turn inhibits the shuttle phenomenon of polar lithium polysulfides, which is highly soluble in electrolyte, based on strong chemical adsorption ability between, and at the same times improves conductivity to deliver high specific capacity at high current densities. As a result, the free-standing electrode with 2.54 mg cm^{-2} sulfur loading exhibited a capacity of about 650 mAh g^{-1} at 5.0 C. By maximizing the advantages of the free-standing electrode, the areal loading of the active material has been increased to 6.11 mg cm^{-2}, which delivers an areal capacity of 3.1 mAh cm^{-2} at 500th.

629445 Controlling Crystallization for Large Ligand Incorporation into Quasi-2D Perovskite Solar Cells

Abstract:

Organic-Inorganic Hybrid Perovskite materials have swept the field of photovoltaics, offering ideal absorbing layers that feature long carrier lifetime and diffusion lengths, strong photoluminescence, and promising tunability. Furthermore, the solution-processing methods used to make these perovskites ensure that the solar cells will remain low-cost. One of the main issues holding back perovskites from becoming ubiquitous is their stability in how they degrade in the presence of water, thus leading to decreased device performance. One promising method found to improve the stability is to reduce the dimensionality of the perovskites into 2D perovskites. This is achieved by incorporating large organic ammonium ligands that are too large for the Goldschmidt Tolerance factor, thus reducing dimensionality of the perovskite. While these 2D variants have improved moisture stability over $MAPbI_3$, the incorporation of organic ligands introduces electronically-insulating layers that inhibit charge transport between the conducting inorganic slabs. Ordinarily this would not be very impactful, except that by general methods used to produce solar perovskite films prefer an in-plane alignment that causes there to be no path for photo-generated carriers to be extracted out of the device. To this end, vertical orientation or "out-of-plane," orientation is critical for the highest efficiency devices. Several efforts have been made to determine the mechanism of growth to promote vertical-orientation in these quasi-2D perovskites involving methods such as hot casting, additive engineering, and other fabrication engineering methods. While devices with substantial power conversion efficiencies (PCEs) have been made over the past few years, the vast majority of the devices with PCEs over 10% feature small organic ligands such as butylammonium (BA) and phenylethylammonium (PEA), which severely limit the potential of 2D perovskites as absorbing layers. The goal now is to extend high efficiency 2D perovskite devices to longer ligands with multiple conjugated groups. It is thought that many of the current techniques emphasize two points in forming the perovskite layer: 1) Isolate nucleation on surface of film (liquid/vapor interface) and 2) have slow growth for high quality junctions. When extending these techniques to larger, multi-functional conjugated ligands, the differences in solubilities for precursors is more extreme, requiring tighter control of the nucleation and crystal growth process. Using solvent engineering methods involving more poorly coordinating solvents (with respect to PbI_6^{2-} octahedra) and creating a solubility-gradient at the vapor-liquid interface in thin films with high-vapor pressure cosolvents, we have determined a method to help vertically orient quasi-2D perovskite with large multi-conjugated group ligands, namely $IH_3N(CH_2)_2(SC_4H_2)_2$ aka 2T. The vertical orientation has been characterized via many diffraction methods (XRD, GIWAXS, etc) and shown an improvement of devices over 13% PCE, doubling the efficiency of similarly reported devices in literature. Furthermore, the devices have tremendous moisture stability mostly due to their bulky conjugated ligands, being able to retain their performance better than their BA counterparts, and far better than their 3D perovskite devices. In addition, through this co-solvent engineering method, the trap density of the perovskites has greatly improved in addition to creating pinhole-free films. While creating high-performing devices is important, this study contributes to the field even more by focusing on the understanding the control requirements on quasi-2D perovskite photovoltaics with larger conjugated ligands. This step is critical to the development of perovskites as solar cells because it can be extended to even larger and more complex semi-conduction ligands making the perovskite system more moisture stable, bringing us all one step closer to the realization of affordable, and ubiquitous solar energy.

631307 Evaluation of Life Cycle Environmental Impacts of Chemical Precursors Used in Perovskite Photovoltaics Manufacturing.

Abstract:

Lead halide perovskite photovoltaics (LHP PVs) have rapidly reached a record efficiency of 25.2%, positioning them to compete with current PV technologies developed over decades. Combined with their low fabrication costs, earth-abundant materials, and superior optoelectronic performance, LHP PVs are on track for near-future commercialization. Key challenges exist today to improve the module operational lifetime, scale up manufacturing, and derisk the presence of toxic lead. Devices fabricated with a mixed cation perovskite absorber layer such as cesium/methylammonium/formamidinium ($Cs_xMA_yFA_{1-x-y}PbI_3$) show longer operational stability and higher efficiency than the most-studied $MAPbI_3$. However, earlier life cycle assessment (LCA) literature based on lab scale analysis reported an outsized environmental impact for FA, raising concerns over commercializing high-performing compositions.

In this talk, we will present a comprehensive process-based LCA study that uses updated data sources, process scale-up concepts, and sensitivity analysis to build ex ante commercial-scale life cycle inventory models for perovskite precursors that can inform industrial manufacturing choices with more transparent and robust environmental analysis. We have found that the impacts between Cs/FA/MA across climate change, cumulative energy demand, and human toxicity categories are all similar to each other. Resource efficiency calculations reveal that cesium supply is enough for near future perovskite PV deployment. Process contribution analysis reveals that reducing synthesis solvent usage, for instance by adding a solvent recovery step, may reduce impacts. However, the impacts for precursors, including lead iodide, are already 1,000 times less than those of other module components like solar glass or flexible substrates. Therefore, selection of perovskite composition can be based on PV efficiency and operational stability, without additional constraints of environmental impact.

626032 Direct Solution Deposition of Metal Selenide Semiconductors Using Novel Metal-Selenium Complexes

Abstract:

In the field of semiconductor materials, considerable research effort has been dedicated to the development of solution-deposition techniques due to the potential of these methods to reduce costs while increasing manufacturing throughput compared to analogous vacuum deposition techniques. In the vein of solution-processed thin-film solar cells, $Cu(In,Ga)(S,Se)_2$ has received significant attention due its combination of high efficiencies and stability. As with many other solution-deposited metal chalcogenide semiconductors, hydrazine-based molecular precursors have set the bar by producing solar cells with efficiencies exceeding 17%.[1] The success of hydrazine-based methods in solution-deposited $Cu(In,Ga)(S,Se)_2$ and other metal chalcogenides can be explained by understanding its reactive-dissolution chemistry. Hydrazine, with the addition of excess chalcogen, is capable of dissolving a range of metal chalcogenides, forming a number of hydrazinium thiometallates and hydrazinium selenometallates.[2] These soluble species will then decompose upon heating into the desired metal chalcogenides and other volatile products that are easily removed from the film. However, hydrazine is both explosive and highly toxic, potentially hindering the scale-up of this method.

While other chemistries have been developed to replace hydrazine in the solution deposition of metal sulfides, limited success has been seen in developing a molecular precursor approach to metal selenides.[3] This is in large part due to the difficulty in producing soluble and stable selenium containing species. One promising route to overcome this limitation is through the use of the amine-thiol reactive solvent system. The amine-thiol system has emerged as one of the most promising alternatives to hydrazine, producing $Cu(In,Ga)(S,Se)_2$ devices with efficiencies above 15%.[4] Not only can this reactive solvent system dissolve metals and metal sulfides, but the ability of this system to dissolve selenium and metal selenides provides an opportunity to match hydrazine in producing metal selenide films.[5] However, one significant challenge to this prospect is that the metal-thiolates produced in this dissolution can easily decompose into metal sulfides upon heating, often leading to significant sulfur content in the precursor film.[6]

In this research, we will discuss our investigations into the metal-chalcogenide complexes that are formed with the amine-thiol solvent system through the use of ^{1}H-NMR, Raman, and ESI-MS. We will then present how an understanding of this chemistry allows for the synthesis of novel metal-selenium complexes that can be used for the direct deposition of $Cu(In,Ga)Se_2$ films. We will further discuss the role of sulfur and selenium in the precursors film and how the chalcogen species alters the grain growth mechanism, contributing to the formation of the notorious "fine-grain" layer that is often observed during selenization of sulfide precursor films. Finally, we will show the broader applicability of these techniques by extending these methods to the direct deposition of pure selenide $Cu_2ZnSnSe_4$.

In conclusion, this research focuses on furthering our understanding of soluble metal complexes, using that knowledge to alter their molecular structure and produce novel metal-selenium complexes, and applying these complexes to the solution-processed fabrication of $Cu(In,Ga)Se_2$ thin films. While the primary focus of this work was directed towards $Cu(In,Ga)Se_2$, we have also shown that these methods have broader applicability to other solution processed metal selenides.

(1) Zhang, T.; Yang, Y.; Liu, D.; Tse, S. C.; Cao, W.; Feng, Z.; Chen, S.; Qian, L. High Efficiency Solution-Processed Thin-Film $Cu(In,Ga)(Se,S)_2$ Solar Cells. *Energy Environ. Sci* **2016**, *9*, 3674. https://doi.org/10.1039/c6ee02352e.

(2) Chung, C.-H.; Li, S.-H.; Lei, B.; Yang, W.; Hou, W. W.; Bob, B.; Yang, Y. Identification of the Molecular Precursors for Hydrazine Solution Processed $CuIn(Se,S)_2$ Films and Their Interactions. *Chem. Mater* **2011**, *23*, 964–969. https://doi.org/10.1021/cm103258u.

(3) Clark, J. A.; Murray, A.; Lee, J.-M.; Autrey, T. S.; Collord, A. D.; Hillhouse, H. W. Complexation Chemistry in N,N-Dimethylformamide-Based Molecular Inks for Chalcogenide Semiconductors and Photovoltaic Devices. *J. Am. Chem. Soc.* **2018**, *141*, 298–308. https://doi.org/10.1021/jacs.8b09966.

(4) Yuan, S.; Wang, X.; Zhao, Y.; Chang, Q.; Xu, Z.; Kong, J.; Wu, S. Solution Processed Cu(In,Ga) $(S,Se)_2$ Solar Cells with 15.25% Efficiency by Surface Sulfurization. *Appl. Energy Mater.* **2020**, *3*, 6785–6792. https://doi.org/10.1021/acsaem.0c00917.

(5) Zhang, R.; Cho, S.; Lim, D. G.; Hu, X.; Stach, E. A.; Handwerker, C. A.; Agrawal, R. Metal–Metal Chalcogenide Molecular Precursors to Binary, Ternary, and Quaternary Metal Chalcogenide Thin Films for Electronic Devices. *Chem. Commun.* **2016**, *52* (28), 5007–5010. https://doi.org/10.1039/C5CC09915C.

(6) Deshmukh, S. D.; Ellis, R. G.; Sutandar, D. S.; Rokke, D. J.; Agrawal, R. Versatile Colloidal Syntheses of Metal Chalcogenide Nanoparticles from Elemental Precursors Using Amine-Thiol Chemistry. *Chem. Mater.* **2019**, *31* (21), 9087–9097. https://doi.org/10.1021/acs.chemmater.9b03401.

632395 Double Cation Substitution of CZTS Chalcogenide Semiconductors for Improved Device Performance

Abstract:

Solution processed thin-film photovoltaic (PV) technology is poised to be the future of manufacturing standard for commercially available solar energy modules due to the substantially low estimated capital expenditure (CAPEX) required. Over the decades, the concept of high efficiency CZTSSe has enticed researchers but always eluded their grasps. This report explores improving the device performance metrics of CZTSSe with the goal of improving the PCE in hopes that it become industrially disruptive to the currently adopted technologies. The approach taken in this research was to conduct double cation alloying with elements shown in literature to have beneficial impacts for CZTSSe device performance with the hypothesis that the combination would be complementary. First, we investigated the combination of small stoichiometric substitutions of Cu with Ag and Zn with Mn so that the resulting stoichiometry would resemble (Agx,Cu1-x)2(MnyZn1-y)Sn(S,Se)4. Second we investigated the double cation substitution of Cu with Ag and Sn with Ge so that the already proven effects of Ge incorporation could be correlated with the effects of Ag. Collectively, the work included in this report demonstrates considerable progress exploring uncharted avenues toward increasing the performance of CZTSSe. By attempting a new approach to understanding optoelectronic performance and reliability in CZTSSe the goal of high efficiency CZTSSe may soon be attainable.

628716 Optoelectronic Characterization of Silver-Doped Cu_3AsS_4 for Photovoltaic Applications

Abstract:

Enargite (Cu_3AsS_4) is an emerging photovoltaic material, consisting of earth abundant elements, with a bandgap in the vicinity of 1.41 eV. Its constituent atoms, Cu, As, and S, have differing ionic radii of 91, 60, and 170 pm respectively, reducing the likelihood of cation disorder. Theoretical studies conducted on this material have suggested that enargite is expected to have high carrier mobilities, high defect tolerance with shallow defects, and benign grain boundaries.[1] Subsequent experimental work on enargite, Cu_3AsS_4, from our group has demonstrated that enargite films have a Shockley-Read-Hall, or defect assisted, recombination lifetime on the timescale of nanoseconds, relatively shallow defect energies, and carrier concentrations on the order of 10^{15} cm^{-3}.[2,3] However, carbonaceous fine grain layers, secondary phases, and pinholes may be part of the reason that power conversion efficiencies for enargite-based photovoltaic devices remained low. In this work, silver doping of Cu_3AsS_4 films facilitated by a solution-processed route using amine-thiol based molecular precursor inks is shown to improve film morphology and the optoelectronic properties of enargite-based photovoltaic devices.

The amine-thiol solvent system is capable of dissolving a variety of metal chalcogenides for the fabrication of thin films and nanoparticles.[4,5] Arsenic sulfides and copper sulfides can undergo reactive dissolution in amine-thiol solvents to form complexes that can decompose at temperatures below 400°C. These molecular precursor solutions can be combined into inks that can directly be coated onto glass substrates to create tetragonal luzonite (Cu_3AsS_4) films. These nanocrystalline films can then be coarsened in an arsenic-sulfide atmosphere at 425°C to produce an orthogonal enargite (Cu_3AsS_4) film with 0.5 μm sized grains. Unfortunately, devices made using these enargite films demonstrate low shunt resistances, poor film morphology, and low external quantum efficiency (EQE).

Photoluminescence (PL) characterization on the coarsened enargite film shows a large peak at 0.9 eV. As this peak is roughly 0.5 eV from the band edge, it would be considered a mid-gap defect and is likely a large contributor to the poor performance of enargite-based devices. Computational analysis done on the defect properties of enargite shows that S_{Cu} and S_{As} anti-site defects would have energies near 0.5 eV from the enargite valence band and could be the cause for the large PL peak seen at 0.9 eV.[6] In hopes of increasing shunt resistances, improving film morphology, and improving optoelectronic properties, inspiration was taken from ongoing research on silver doping of $Cu_2ZnSnSe_4$ (CZTSe) and $Cu(In,Ga)Se_2$ (CIGSe) thin film solar cells.

In both CZTSe and CIGSe films, silver doping has been shown to decrease the impact of copper-based defects by modifying their formation energy and their location in the absorber band gap.[7,8] Furthermore, silver alloying has been shown to increase grain size and reduce structural defects, which may help improve the morphology of enargite films and increase shunt resistances. The amine-thiol solvent system allows for facile incorporation of varying concentrations of Ag_2S into coating inks for the creation of silver doped enargite films. Films with a silver content ranging from 0 to 0.1 Ag/(Cu+Ag) were created and analyzed through various optoelectronic characterization techniques.

Silver incorporation increases the average size of grains in coarsened enargite films from 560 nm to 1.11 μm. The overall morphology of the film also improves with no visible pinholes in SEM micrographs. Kelvin-probe force microscopy (KPFM) shows that the grain boundaries of enargite films at all levels of silver doping are negatively charged (having a positive contact potential difference) in comparison to the bulk grains, suggesting that grain boundaries will not act as shunt pathways for minority carriers in the device. Time-resolved photoluminescence shows a doubling in minority carrier lifetimes from 0.23 ns for 0

Ag/(Cu+Ag) to 0.46 ns for 0.1 Ag/(Cu+Ag). The maximum EQE of enargite increases with increasing silver content from 1% to over 8% at 0.5 Ag/(Cu+Ag). The bandgap of enargite was extracted from the EQE spectra, and similar to what occurs in silver doped CZTSe and CIGSe, silver doping increases the band gap of enargite from 1.43 eV for no silver to 1.50 eV for 0.1 Ag/(Cu+Ag). Photoluminescence measurements on silver-doped enargite films show that the relative intensity of the primary band-to-band recombination peak decreases with increasing silver content. This may be due to silver-based defects in the films leading to an increase in non-radiative recombination. The relative intensity of the photoluminescence peak at 0.9 eV does not change with silver doping, suggesting that this peak may not be related to a copper-based defect.

Silver-doped enargite devices show a maximum champion efficiency of 0.6% at 0.02 Ag/(Cu+Ag). Improved film morphology and increased grain size leads to an initial increase in shunt resistance and J_{sc}. However, likely due to increases in bandgap energies with increasing silver content, J_{sc} decreases after 0.05 Ag/(Cu+Ag) from a maximum of 1.9 mA/cm^2. Increasing bandgap energies also contributes to the initial increase in $V_{oc.}$ However, increases in non-radiative recombination at higher levels of silver doping causes V_{oc} to decrease from a maximum of 170 mV at 0.02 Ag/(Cu+Ag) to 80 mV at 0.1 Ag/(Cu+Ag).

This work demonstrates that silver-doping can be employed on emerging photovoltaic materials and may produce similar effects as is seen in AZTSe and CIGSe, including increased grain size, increased bandgap, and improved minority carrier lifetimes. In enargite films, the ideal amount of silver doping is around 0.02 Ag/(Cu+Ag) as it maximizes improvements in film morphology and minority carrier lifetimes while minimizing the drawbacks involved with enlarged bandgap energies and increased silver-based defects.

[1]Wallace, S. K., Svane, K. L., Huhn, W. P., Zhu, T., Mitzi, D. B., Blum, V., Walsh, A., "Candidate photoferroic absorber materials for thin-film solar cells from naturally occurring minerals: enargite, stephanite, and bournonite," S*ustainable Energy and Fuels,* **1**, 1339, (2017).

[2]McClary, S. A., Andler, J., Handwerker, C. A., Agrawal, R., "Solution-processed copper arsenic sulfide thin films for photovoltaic applications," *Journal of Materials Chemistry C,* **5**, 28, pp. 6913-6916 (2017).

[3]McClary, S. A., Taheri, M. M., Blach, D. D., Pradhan, A. A., Li, S., Huang, L., Baxter, J. B., Agrawal, R., "Nanosecond carrier lifetimes in solution-processed enargite (Cu_3AsS_4) thin films," *Applied Physics Letters*, **117**, 16, pp. 162102 (2020).

[4]McCarthy, C. L, Brutchey, C. L., "Solution processing of chalcogenide materials using thiol-amine "alkahest" solvent systems," *Chemical Communications*, **53**, pp. 4888-4902 (2017).

[5]Zhao, X., Deshmukh, S. D., Rokke, D. J., Zhang, G., Wu, Z., Miller, J. T., Agrawal, R., "Investigating Chemistry of Metal Dissolution in Amine-Thiol Mixtures and Exploiting It Toward Benign Ink Formulation for Metal Chalcogenide Thin Films," *Chemistry of Materials,* **31**, 15, pp. 5674-5682 (2019).

[6]McClary, S. A., Yao, C., Rokke, D. J., Pradhan, A. A., Andler, J., Yan, Y., Agrawal, R., "A combined Experimental and Theoretical Study of the Defect Properties of Enargite Cu_3AsS_4," *In Prep.*

[7]Hages, C. J., Koeper, M. J., Agrawal, R., "Optoelectronic and material properties of nanocrystal-based CZTSe absorbers with Ag-alloying," *Solar Energy Materials and Solar Cells,* **145**, 3, pp. 342-348 (2016).

[8]Zhao, Y., Yuan, S., Kou, D., Zhou, Z., Wang, X., Xiao, H., Deng, Y., Cui, C., Chang, Q., Wu, S., "High Efficiency CIGS Solar Cells by Bulk Defect Passivation through Ag Substitution Strategy," *ACS Applied Materials & Interfaces,* **12**, 11, pp. 12717-12726 (2020).

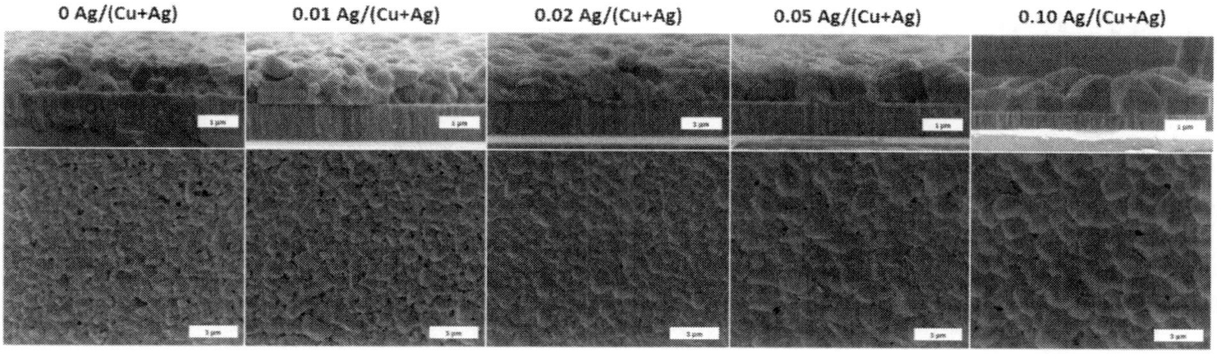
0 Ag/(Cu+Ag)
0.01 Ag/(Cu+Ag)
0.02 Ag/(Cu+Ag)
0.05 Ag/(Cu+Ag)
0.10 Ag/(Cu+Ag)

628423 Deeply Rechargeable Zinc Anode Materials for Ultra-Safe High-Energy Rechargeable Batteries

Abstract:

Metallic zinc as a rechargeable anode material for aqueous batteries has gained tremendous attention with merits of intrinsic safety, low cost, and high theoretical volumetric capacity (5,854 mAh/cm3). Among zinc-based batteries, Zn-air batteries are promising with the highest theoretical volumetric energy density (4,931 Wh/L). Rechargeable zinc anode has recently achieved progress in neutral electrolytes, yet developed slowly in alkaline electrolytes, which are kinetically favorable for air cathodes. Passivation, dissolution, and hydrogen evolution are three main reasons for irreversibility of zinc anodes in alkaline electrolytes, which limits the rechargeability and usable energy density. In this talk, I will present our recent works on using nanoscale material design to overcome passivation, dissolution, and hydrogen evolution issues of zinc anode, towards a deeply rechargeable zinc-based battery. I will also introduce the battery-gas chromatography quantitative analysis, as well as in situ microscopy methodologies we have developed, to quantify gas evolution side reaction, as well as visualize the reaction on electrodes during operation.

632641 Ni-Fe Layered Double Hydroxides and Ni-Sulfide Electrocatalyst Foams for Electrochemical Energy Storage and Conversion Devices

Abstract:

Ni-Fe Layered Double Hydroxides and Ni-Sulfide Electrocatalyst Foams for Electrochemical Energy Storage and Conversion Devices

Enoch A. Nagelli[1,2*], Caspar Yi[1,2], F. John Burpo[1,2], Sean P. Rogers[1], Matthew DiBiase[1], Jiangtian Li[3], Rongzhong Jiang[3], Deryn Chu[3*]

[1]Department of Chemistry & Life Science, Chemical Engineering Program, United States Military Academy, West Point, New York 10996

[2]Photonics Research Center, United States Military Academy, United States Military Academy, West Point, New York 10996

[3]U.S. Army Combat Capabilities Development Command, Army Research Laboratory, Sensors and Electron Directorate, 2800 Powder Mill Rd., Adelphi, MD 20783-1107

Electronic conductivity of porous electrodes by Faradaic and non-Faradaic charge transfer processes impact the energy and power densities and overall performance of electrochemical energy storage and conversion devices. Advances in three-dimensional (3D) porous electrode materials design offers a selection of cost-effective non-precious metal porous electrodes such as nickel for electrochemical energy applications. These alternative catalysts include nickel foams with high surface area and surface-to-volume ratio are promising materials to reduce kinetic losses and enhance mass transfer rates, respectively. Electrochemical devices such as fuel cells rely on electrolysis to produce hydrogen and oxygen gases with expensive porous catalysts. Supercapacitor performance is impacted by slow ion diffusion and lower conductivity through the porous electrodes. Transition metal layered hydroxides have been demonstrated to be conductive and feasible supercapacitor electrode materials with enhanced double-layer capacitance, as well as high performing fuel cell electrocatalysts [1-3]. In this study, we investigate the role of surface morphology, elemental composition, and crystal structure of Ni-Fe layered double hydroxide (LDH) and Ni-S based foams produced from electrodeposition and hydrothermal synthesis techniques on electrocatalysis for electrochemical anodic and cathodic water splitting reactions (hydrogen evolution reaction/HER and oxygen evolution reaction/OER). Electrocatalytic performance of the Ni-Fe LDH and Ni-S based 3D foams was evaluated using linear sweep voltammetry to determine the minimum overpotentials of the catalyst necessary to drive HER and OER. Moreover, galvanostatic charge-discharge cycling of the NiFe-LDH and Ni-S foams was conducted to determine the capacitance and overall performance as supercapacitor electrodes. Electrochemical impedance spectroscopy was used to determine the electronic and ionic conductivity of the hierarchical porous foams. Material characterization techniques such as scanning electron microscopy (SEM), energy dispersive x-ray spectroscopy (EDX), and x-ray diffractometry (XRD) were used to characterize the surface morphology, surface elemental composition, and crystallinity, respectively.

KEYWORDS: Fuel Cells, Supercapacitors, 3D Foams, Electrocatalysis, Oxygen Evolution Reaction,

Hydrogen Evolution Reaction, Electrochemical Engineering, Energy Storage

Corresponding Authors: Enoch A. Nagelli, Email: enoch.nagelli@westpoint.edu Deryn Chu, Email: deryn.d.chu.civ@mail.mil

REFERENCES

1. Sim et al. *Mater. Chem. A*, **2018**, 6, 1504-1512.,
2. Xiong et al. *Funct. Mater.*, **2021**, 31, 2009580.
3. Yu et al. *Rep.* **2018**, 8, 5246.

629865 Rethinking Convective Chromatographic Separations

Abstract:

The use of membrane supports with immobilized protein A has not lead to a technological breakthrough as it was expected when the first protein A affinity membranes appeared on the market in the late 80's/early 90's. Their excellent flow properties were not sufficient to compensate the low binding capacity with respect to chromatographic beads for protein capture applications. Nevertheless, the development and booming of single use technology explains the success of membrane adsorbers for impurity removal in biopharmaceutical manufacturing.

Besides the widespread application of membrane chromatography in polishing steps, there is still a large and growing research interest in its implementation for capture and purification of high added value products such as antibodies and other therapeutic molecules. This research effort led to a remarkable progress in developing highly selective and efficient stationary phases. Indeed, the binding capacity of protein A immobilized on an optimized membrane base material, with respect to commercially available protein A membranes, could be increased about seven-fold that demonstrates the tremendous effect that the membrane support has on binding capacity. However, these values are still too low with respect of the capacity of chromatography beads, even if competitive binding capacities have been obtained with nonwoven ion-exchange membrane adsorbers.

Moreover, continuous biomanufacturing will underline the limits of the conventional chromatographic process based on packed bed columns and highlight the need of a novel chromatographic process. In this work, innovative supports for convective chromatography produced with 3D printing will be presented. These new stationary phases, inspired by nature, have the potential to dramatically change chromatographic separations.

632551 Covalent Organic Framework As an Ion Exchange Material for Adsorptive Separation of Biomolecules

Abstract:

In this work, the use of covalent organic frameworks (COFs) in biomolecule separation has been explored. We examined the ionic COF Py-BPy^{2+}-COF as an ion exchange material for protein and amino acid separation. After characterizing the physical and chemical properties of the synthesized COF with a variety of techniques such as Fourier transform infrared analysis, powder X-ray diffraction, X-ray photoelectron spectroscopy, and transmission electron microscopy, binding experiments with both large and small biomolecules were performed. High adsorption capacities of amino acids with different hydrophobicity and charge, as well as proteins of different isoelectric points and molecular weights were determined in batch equilibrium experiments. Desorption experiments with model proteins mixture demonstrated an ability to successfully separate one protein from another with selectivity hypothesized to be a combination of isoelectric point, hydrophobicity, and ability to penetrate the crystalline material. The results demonstrated that Py-BPy^{2+}-COF can be exploited as a robust crystalline anion exchange biomolecule separation material.

632787 Scalable Synthesis of Nanoporous Atomically Thin Graphene Membranes for Dialysis and Molecular Separations Via Facile Iso-Propanol-Assisted Hot Lamination

Abstract:

Scalable graphene synthesis and facile large-area membrane fabrication are imperative to advancenanoporous atomically thin membranes (NATMs) for molecular separations. Although chemical vapordeposition (CVD) allows for roll-to-roll high-quality monolayer graphene synthesis, facile transfer withatomically clean interfaces to porous supports for large-area NATM fabrication remains extremely challenging.Sacrificial polymer scaffolds commonly used for graphene transfer typically leave polymer residues detrimentalto membrane performance and transfers without polymer scaffolds suffer from low yield resulting in high non-selective leakage through NATMs. Here, we systematically study the factors influencing graphene NATMfabrication and report on a novel roll-to-roll manufacturing compatible isopropanol-assisted hot lamination (IHL)process that enables scalable, facile and clean transfer of CVD graphene on to polycarbonate track etched(PCTE) supports with coverage $\geq$99.2%, without compromising support integrity/porosity. We demonstrate fullyfunctional centimeter-scale graphene NATMs that show ~2-3 orders of magnitude higher permeance andbetter selectivity than commercially available state-of-the-art polymeric dialysis membranes, specifically in the0-1000 Da range. Our work highlights a scalable approach to fabricate graphene NATMs for practicalapplications and is fully compatible with roll-to-roll manufacturing processes.

Cheng et al. Nanoscale. 2021
Cheng et al. Nano Lett. 2020
Kidambi et al. Adv. Mat. 2018
Kidambi et al. Adv. Mat. 2018
Kidambi et al. ACS App. Mat. Int. 2018
Kidambi et al. Adv. Mat. 2017
Kidambi et al. Adv. Mat. 2017
Kidambi et al. Nanoscale 2017

627558 Immiscible Liquid-Coated Materials for Bioseparations

Abstract:

Bioseparations, such as liquid and aerosol filtration, are most often performed with the use of porous materials. However, porous materials can be prone to biofouling and other types of surface adsorption. In this work, we present bio-inspired liquid-coated surfaces as a new approach to separation materials that resist fouling. Liquid-coated filters can be created by immobilizing a water immiscible liquid on the surface of commercially available synthetic filtration membranes. By chemically matching the infusing liquid to the filtration membrane, the stable liquid layer that is formed produces a dynamic, mobile surface that resists the adhesion of contaminants. For water purification, 0.45 µm pore diameter polytetrafluoroethylene (PTFE) and polyvinylidene fluoride (PVDF) filters were tested using omniphobic perfluoropolyether coating liquids. We measured the continuity of the surface liquid layer by testing how easily a water droplet could begin to move the surface, as well as the anti-adhesion properties of the surface liquid layer through quantifying the speed of droplet movement at a static angle of inclination. The results indicate that the PTFE membranes could sustain a more consistent functional liquid layer, with an approximately 75% lower sliding angle and 70% faster droplet movement at a fixed angle than the PVDF. For aerosol filtration, we tested liquid-coated commercial PTFE and HEPA filters for their efficiency in capturing and releasing aerosolized *Escherichia coli* for analysis. We determined that bacteria trapped on the liquid-coated filters could be removed with increased efficiency compared to bare controls. The use of liquid coated materials in bioseparations opens new doors for the creation of dynamic filtration strategies.

623769 Cu-Selective Membrane Adsorbers for Medical Isotope Production

Abstract:

Radiopharmaceuticals offer promising new approaches for imaging and treatment for cancer patients; however, the current supply of these novel drugs is too low to support clinical trials. Copper-67 (Cu-67), can provide treatment (via beta decay) and imaging (via gamma rays). Cu-67 is purified using resin-packed columns, which are known to suffer from diffusion-limited transport. The practical implications of these limitations are (1) low column flow rates, (2) high elution volumes and (3) long purification times—all of which contribute to the bottleneck in the sustained supply of Cu-67. Due to the short half-life (2.58 days) of Cu-67, rapid purification is key for increasing the availability of the isotope for clinic trials and use.

A promising alternative to resin-packed columns are membrane adsorbers, which have seen recent success in the purification of biologics. This contribution describes our efforts to graft glycidyl methacrylate (GMA) from polyvinylidene fluoride (PVDF) membranes through activator generated by electron transfer, a type of, atom transfer radical polymerization (AGET ATRP). After grafting poly(GMA), diamine ligands of varying chain length (putrescine and ethylene diamine) are attached to the brushes through an epoxide ring opening reaction. Membranes are characterized by attenuated total reflection Fourier-Transform infrared spectroscopy throughout the synthesis process (grafting and ring-opening) to support functionalization. Membrane permeability was calculated from pure-water flux experiments in a dead-end filtration cell to be 5166 LMH/bar for unmodified PVDF membranes, 4692-648 LMH/bar for GMA modified membranes, and 4124-175 LMH/bar for the amine functionalized GMA membranes. Membrane morphology and surface chemistry was examined using electron microscopy (SEM) and SEM x-ray photoelectron spectroscopy (XPS). XPS analysis confirms that the Cu catalyst was successfully washed from the membrane after AGET ATRP. Batch adsorption experiments were conducted for Cu and competitor ions (Ni, Zn) in pH 2 HCl solutions. Ion concentrations were measured using an inductively-coupled plasma optical emission spectrophotometer (ICP-OES) and the binding capacity was calculated by modeling equilibrium adsorption data with the Langmuir isotherm. The results of this research are laying the groundwork for implementing membrane absorbers as a new separation material for medical isotope purification.

628735 Tuning Structural Defects on Nominal Single-Layered Graphene Oxide Membrane for Selective Separation of Biomolecules

Abstract:

The advent of 2-dimensional materials provided a unique opportunity to design and synthesize the ideal membrane with ultrathin thickness and precisely controlled molecular-sized pores. Graphene oxide (GO), owing to its water solubility and functionality, has attracted enormous attention for membrane applications. However, standard synthesis of an ideal high-performance membrane is still limited by the thickness of GO membranes, leading to a loss in permeance. Optimizing the deposition methodology of GO flakes on polymeric substrates could address this limitation and enable the synthesis of ultrathin single-layered membranes. In this study, a sequential coating methodology was adopted for depositing single layer GO flakes on a Polycarbonate polymeric support with increments of 10% area-based coating of GO, followed by different time exposure to oxygen plasma in order to form tunable transport pathways on the surface of GO. Also, retention of different single model proteins (BSA, Lysozyme & IgG) and of mixtures of these proteins were measured. In addition, the separation efficiency for similar sized proteins (myoglobin and lysozyme) exhibited separation factors of ~6 with purities of ~92% through the plasma-etched GO membrane. These findings offer new opportunities of using GO flakes to obtain ultrathin membranes on polymeric supports and fabricate tunable GO membranes for possible application in biotechnology industry.

626234 The Development of Nanocarbon Immobilized Membrane for Elimination of Thermophilic Bacteria Via Membrane Distillation

Abstract:

Pathogen contamination in water has always been a threat to public health causing diseases such as dysentery, typhoid and cholera and such outbreaks have led to millions of deaths worldwide. Methods for treating pathogen-contaminated water include boiling, treatment with ozone, chlorine, UV irradiation, and filtration. These methods have their limitations, can be expensive, and can also produce secondary hazardous wastes such as disinfection byproducts that themselves are toxic. Recently, membrane processes which have low cost, smaller footprint, and provide effective treatment options for a variety of pollutants have become popular in water purification.

Membrane distillation (MD) is a separation technique where the driving force is a vapor pressure gradient between a hot feed and a cold permeate. While many microorganisms get killed at higher temperatures, the challenge arises in cases where these microorganisms are stable at high temperatures namely thermophilic bacteria. In this study, carbon nanotube, functionalized carbon nanotube, and graphene oxide immobilized membranes were used in membrane distillation to generate microorganism free water. The nanocarbons degraded the bacterial cell integrity and caused cell death. The pure CNTs showed the highest biocidal activity (96.2%) followed by functionalized CNTs and graphene oxide. The biocidal activity was significantly higher than the plain polymeric membranes.

In general, the biocidal action of nanocarbons typically involves a combination of physical and chemical mechanisms. First, the nanostructured CNTs acted as a mesh to trap the bacterial cells. The presence of van der waals forces facilitated the adhesion of bacterial cells to the surface of CNTs. The sharp structures of CNTs are also known to cause significant structural damage to the cell wall and the membrane of the microorganism. On the other hand, GO exists as layers of nanosheets, and we believe that initially the bacterial cells get trapped within the nanosheets. Bacterial cell death in the case of GO is believed to be caused via oxidative stress. The presence of GO induces the production of reactive oxygen species (ROS) within the bacterial cells, which causes a decrease in membrane fluidity by altering membrane properties, ultimately leading to cell death. The CNT membranes also showed less fouling. While biofilm/bacterial sludge accumulated on the CNT-COOH and GO immobilized membrane surfaces, no biofilm was observed on the CNT membrane. Among the membranes studied, the CNTs appeared to have the highest biocidal effect and lowest biofouling.

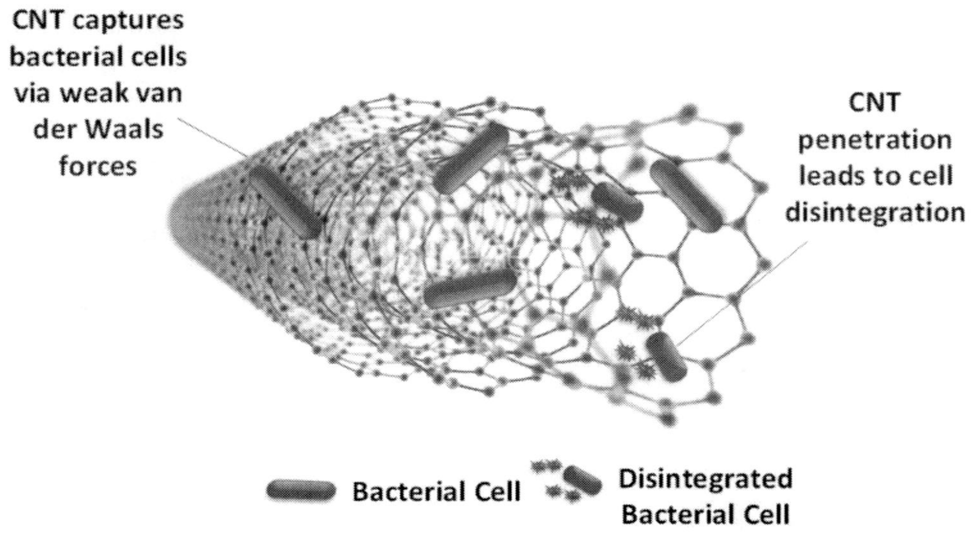
CNT captures bacterial cells via weak van der Waals forces
CNT penetration leads to cell disintegration
Bacterial Cell
Disintegrated Bacterial Cell

632569 Bioinspired Nanoporous Ion Conducting Membranes for Next Generation Batteries

Abstract:

Bio-inspired ion transport membranes have been widely investigated for energy storage applications. High theoretical specific energy density (2600Wh/kg) and high specific capacity (1675mA/g) along with natural abundance and low toxicity of sulfur have been attracting significant attention for development of an alternative battery system to replace traditional lithium ion batteries which suffer from safety and capacity/energy density limitations in various applications. However, challenges such as poly sulfide dissolution and shuttling prevent mass commercialization of metal sulfur batteries. Inspired from biological ion transport mechanisms, we show a practical yet comprehensive approach for development of high-performance metal sulfur batteries. Aramid nanofiber (ANF) based composite ion transport membranes not only prevent dendrite formation but also confine poly sulfides on the cathode side. ANF composite battery separators provide diverse and opposing properties including high mechanical properties, high ionic conductivity and high thermal/chemical stability. Highly selective ion sieving properties of these bio-mimetic separators provide safe and high-performance batteries. Fabrication of such bio-compatible, affordable, flexible and high energy density battery is quite crucial in powering next-generation electronics including but not limited to portable, wearable and implantable biomedical devices.

632179 Tailored Trilayer Separator for Extreme Temperature Lithium-Sulfur Batteries

Abstract:

Since the phenomenal invention of Lithium-ion Batteries (LIBs) in the 1970s, the transformation of human lives has been inconceivable. From the introduction of LIBs for portable electronics, they eventually migrated to other variety of applications such as defense, transportation, and space explorations. With the ever-growing demand comes the need for matching the supply of resources. Thus, newer materials with higher capacities and energy densities and different chemistries are being discovered. High-capacity lithium cathodes such as $LiNiMnCoO_2$, $LiCoO_2$, $LiFePO_4$, and $LiMn_2O_4$ have been identified a few decades back, however, these cathodes have capacities <250 mAh g^{-1}. This barrier prevents LIBs from being used for energy-intense applications such as long-range transportation, grid storage, etc. The high theoretical lithium-sulfur chemistry comes helpful in this dilemma. Sulfur is the lightest element to be used as a cathode consisting of a theoretical capacity of 1672 mAh g^{-1}. Also, sulfur is abundant in nature and economical (~$40 per metric tonne), compared to some transition metals like Co (~$48K per metric tonne), used in LIBs. Before the Li-S chemistry can be commercialized, three fundamental challenges need to be addressed viz., lithium dendrite formation, polysulfide shuttling, and low sulfur conductivity. To suppress polysulfide shuttling and enhance the low conductivity of sulfur, various measures have been applied viz., conductive polymers, inverse vulcanization, mesh carbon, metal, or carbon oxide frameworks. These measures lead to a decrease in energy density along with issues, which may warrant side reactions. For dendrite formations, different electrolyte additives, shutdown separators, etc are used, however, they increase the impedance in the system or are expensive alternatives.

We present the use of a multipurpose trilayer separator for Li-S battery to target all the three fundamental concerns, concurrently. The modified separator consists of polypropylene coated with a polydopamine layer, which is further coated with <10 μm graphene layer. The presence of polydopamine layer acts as a barricade for polysulfide by adsorbing them preferably due to the enhanced interactions. The presence of graphene layer helps in improving the conductivity of the carbon-sulfur cathode. The performance of LiS batteries was compared with conventional separators and trilayered ones for rate studies and long-term cycling studies. There was quite a difference in the two systems of cells with different separators. Trilayer separator at 0.1C. 0.2C, 0.5C, 1C, 2C, 3C and 4C exhibited capacities of 925, 833, 644, 480, 326, 260, and 220 mAh g^{-1}, respectively with high capacity retention of >95%. Equally, the battery performance depends on the choice of electrolytes and additives. Usage of 1M LiTFSI in 1:1 (v/v) 1, 3-Dioxolane (DOL): 1, 2-Dimethoxyethane (DME) with additives, provided remarkable results at low and high-temperature ranges. These solvents have a very low melting point, which enables them to work at much lower temperatures. At 0 °C, the cell with the modified separator yielded about 350 mAh g^{-1} capacity at 0.5C over 200 cycles. The performance of the cells was stellar even at 50 °C with 500 mAh g^{-1} capacity at 0.5C rate after 100 cycles. The safety characteristics of the systems were compared using Multiple module calorimetry for functional coin cells, which provided the thermal heat signature. The technology is to be scaled up in the Pouch-cell configurations and checked for their performance. These high-performance batteries presented here can be of immense value for various space and defense applications, where drastic conditions affect the battery performance.

627442 Dual Pseudocapacitive Oxides Accelerate Kinetics of Sulfur Intermediates in Lithium-Sulfur Batteries

Abstract:

The sluggish electrochemical kinetics of sulfur species has impeded the wide adoption of lithium-sulfur battery, which is one of the most promising candidates for next-generation energy storage system. Here, we present the electronic and geometric structures of all possible sulfur species and construct an electronic energy diagram to unveil their reaction pathways in batteries, as well as the molecular origin of their sluggish kinetics. By decoupling the contradictory requirements of accelerating charging and discharging processes, we select two pseudocapacitive oxides to enable the efficient transport of electron/Li^+ to and from sulfur intermediates respectively. After incorporating $Li_xNb_2O_5/Nb_2O_5$ as electron-ion source and MnO_2/Li_yMnO_2 as electron-ion drain, the electrochemical kinetics of sulfur cathode is significantly accelerated by a factor of 5 times and 1.8 times, respectively. This strategy, which couples a fast-electrochemical reaction with a spontaneous chemical reaction to bypass a slow-electrochemical reaction pathway, offers a solution to accelerate an electrochemical reaction, providing new perspectives for the development of high-energy battery systems.

631178 Molecular-Level Characterization of the Electrode-Electrolyte Interfaces in Li Batteries

Abstract:

Most next generation Li battery materials suffer from severe interfacial instabilities. Here, we leverage our unique abilities in solid-state NMR and MRI to characterize, at the molecular-level, the composition, arrangement, and dynamics of the solid electrolyte interphases that form on the surface of emerging anode and cathode technologies for Li batteries. We find that interphase stability (e.g. solubility) and composition (e.g. high conductivity compounds) are paramount to controlling high Coulombic efficiencies and achieving high capacity retentions for a range of electrolyte formulations and cell configurations. Insight from this work provides some of the first atomistic descriptors of electrode-electrolyte interfacial reactions that enable safe and efficient battery operation.

628585 Detecting the Onset of Li Plating during Fast Charging of Li-Ion Batteries Using *Operando E*lectrochemical Impedance Spectroscopy

Abstract:

To promote the continued growth of the electric vehicle (EV) market—as well as to improve the performance of other battery-powered technologies such as phones, laptops, and drones—Li-ion battery charging times need to be significantly reduced. At higher charging rates, Li metal may plate on the graphite anode, which is a possible safety concern and leads to decreased cycle life of the battery.[1] There is clearly a need to develop a consistently reliable electrochemical technique to detect the onset of Li plating during active battery operation..[1,2] Previous work has included techniques that interrogate Li plating after charging, including dV/dQ analysis of discharge profiles after fast charging,[3,4] impedance analysis after fast charging[5,6] and during pseudo-operando measurements,[7] dV/dt analysis of voltage relaxation profiles after fast charging,[5,7] three electrode graphite potential monitoring,[4] and simple optical inspection after many fast charging cycles, among other techniques.[1] Our group has also developed ex situ mass spectrometry titrations to detect and quantify electrically isolated Li deposits,[8] as well as an improved dV/dt technique to detect the onset of Li plating after fast charging.[9]

In this work, we use impedance spectroscopy—and specifically the Distribution of Relaxation Times (DRT) analysis method[10]—to detect in operando the onset of Li plating during extreme fast charging. There is a change in the graphite interfacial impedance upon Li plating on the graphite surface. Coupled with both differential voltage relaxation and mass spectrometry titrations for cross-validation, we can reliably detect the Li plating onset graphite state of charge (SOC) for rates as high as 6C at room temperature and for different graphite loadings with <0.5% graphite capacity sensitivity. This study was primarily carried out using three-electrode cell measurements, but we will also show a proof-of-concept for the application of this technique to more commercially relevant two-electrode cells. We will also discuss possible physical explanations for the observed impedance behavior. This capability to detect the onset of Li plating allows for the safe implementation of fast charging protocols, and represents, to the best of our knowledge, the first operando electrochemical technique for Li plating detection.

References

1. P. P. Paul et al., *Adv. Energy Mater.*, 2100372 (2021).

2. T. Waldmann, B.-I. Hogg, and M. Wohlfahrt-Mehrens, *Journal of Power Sources*, **384**, 107–124 (2018).

3. K. G. Gallagher, D. W. Dees, A. N. Jansen, D. P. Abraham, and S.-H. Kang, *J. Electrochem. Soc.*, **159**, A2029–A2037 (2012).

4. C. Uhlmann, J. Illig, M. Ender, R. Schuster, and E. Ivers-Tiffée, *Journal of Power Sources*, **279**, 428–438 (2015).

5. S. Schindler, M. Bauer, M. Petzl, and M. A. Danzer, *Journal of Power Sources*, **304**, 170–180 (2016).

6. X. Chen, L. Li, M. Liu, T. Huang, and A. Yu, *Journal of Power Sources*, **496**, 229867 (2021).

7. U. R. Koleti, T. Q. Dinh, and J. Marco, *Journal of Power Sources*, **451**, 227798 (2020).

8. E. J. McShane et al., *ACS Energy Lett.*, **5**, 2045–2051 (2020).

9. Z. M. Konz, E. J. McShane, and B. D. McCloskey, *ACS Energy Lett.*, 1750–1757 (2020).

10. E. Ivers-Tiffée and A. Weber, *J. Ceram. Soc. Japan*, **125**, 193–201 (2017).

631020 Catalyst Design for Metal Air Batteries Utilizing a Four-Electron Oxidation and Reduction of Metal Oxides

Abstract:

While lithium peroxide (Li_2O_2) is typically the primary discharge product in a Li-O_2 battery, lithium oxide (Li_2O) poses several significant advantages over lithium peroxide including reduced reactivity with organic solvents and an energy density rivaling that of fossil fuels. Recently, a high-energy-density lithium-oxygen battery using a nick-oxide catalyst and lithium nitrate molten salt electrolyte at T=150 °C has been demonstrated to reversibly oxidize and reduce lithium oxide. Additionally, with the advent of molten salt electrolytes for lithium-air batteries, electrocatalyst design for metal-air systems becomes a promising avenue for lowering reaction overpotentials, because Li_2O, a possible reaction product, has some solubility in these electrolytes. However, detailed mechanisms underlying the high overpotentials are currently not understood in these systems. With fundamental understanding of reaction mechanisms and energetics, it may be possible to design catalysts for a very efficient charge and discharge.

In this work, we systematically investigate the effect of M=Li, Na, K and hydrogen on the reaction thermodynamics and kinetics of M-OER in a metal-oxide based metal-air battery using Density Functional Theory (DFT). Developing the concept of the bulk or "intrinsic" free energy diagram, we show that the bulk thermodynamics of Li-OER are inherently facile compared to Na-OER and K-OER, with intrinsic efficiency rivaling that of catalyzed H-OER *even before catalyst design*. We additionally study the mechanism and kinetics for M-OER on M_2O_2 (0001) surfaces, relevant in electrolytes where M_2O and M_2O_2 are insoluble, finding that the bulk free energy diagram is strongly predictive of reaction energetics due to the "bulk-like" environment of the active site. Finally, we explore Li-OER on NiO (100), a weakly binding, non-polar catalyst, and NiO (111), a strongly binging, polar catalyst, in order to develop catalyst design rules for systems in which M_2O and M_2O_2 have some solubility (i.e. molten salts). We find that both catalysts form monolayers or overlayers of Li_xO_y, and are thus able to achieve similarly low overpotentials as the Li_2O_2 (0001) surface. Our results suggest that the most important design considerations for M-OER electrocatalysts are (1) the lattice matching of M_2O_2 (0001) with the support and (2) conductivity, rather than binding energy.
This work was funded by a DOE Office of Science Graduate Student Research (SCGSR) Program award.

629901 Solvation Effects on Lithium Ion Transport and Reaction on Lithium Metal Anodes

Abstract:

Chemistry and structure of electrolytes define characteristic behaviors on cation diffusion, desolvation, and deposition events that influence redox reactions at electrochemical interfaces. Here we discuss the effects of solvent properties such as reduction potential on the barriers for ion transport and their correlation to reorganization of molecules in the primary solvation shell during their journey from bulk to surface. Closer to the surface, we investigate the electron transfer effect on changing the solvation forces, thus transforming solvation into a desolvation process. We study these phenomena using constrained ab initio molecular dynamics within the Blue Moon ensemble along with separate density functional theory analysis of solvation thermodynamics. Our work is relevant to the ion diffusion and deposition on Li metal anodes, but also to many relevant electrochemical interfaces in electrochemical and bioelectrochemical processes.

AUTHOR INDEX